TECHNOLOGY GUIDE
FOR

UNDERSTANDABLE STATISTICS
SIXTH EDITION
BRASE/BRASE

-FEATURING-

TI-83 GRAPHICS CALCULATOR
COMPUTERSTAT VERSION 5
WINDOWS, MACINTOSH, DOS
MINITAB
WINDOWS, RELEASE 12

CHARLES HENRY BRASE
REGIS UNIVERSITY

CORRINNE PELLILLO BRASE
ARAPAHOE COMMUNITY COLLEGE

HOUGHTON MIFFLIN COMPANY BOSTON NEW YORK

Editor-in-Chief: Charles Hartford
Associate Editor: Mary Beckwith
Editorial Assistant: Kathy Yoon
Senior Manufacturing Coordinator: Sally Culler
Marketing Manager: Ros Kane

Printed in the U.S.A.

Technology Guide ISBN: 0-395-90772-1

Technology Guide with ComputerStat for MS-DOS disk ISBN: 0-395-93033-2

Technology Guide with ComputerStat for Windows disk ISBN: 0-395-93034-0

Technology Guide with ComputerStat for Macintosh disk ISBN: 0-395-93035-9

23456789-PO-02 01 00

Trademarks Used

MINITAB is a registered trademark of Minitab, Inc.

IBM is a registered trademark Of International Business Machines, Inc.

Macintosh is a registered trademark of Apple Computer, Inc.

MSDOS is a registered trademark of Microsoft Corporation

Windows is a trademark of Microsoft Corporation

Additional Copyright Notice for ComputerStat Version 5 MSDOS Software

Preface

The use of appropriate computing technology can greatly enhance a student's learning experience in statistics. This guide provides basic instruction, examples, and lab activities for three technologies:

Texas Instruments TI-83 Graphics Calculator

ComputerStat Version 5 for Windows & Macintosh (also available for in DOS platform)

MINITAB (Windows, Release 12 Student Edition)

The Guide is divided into three self contained parts, one for each technology. Lab activities coordinate with the text *Understandable Statistics* fifth edition by Brase and Brase. There is also a table to coordinate the chapters of the Technology Guide to the text *Understanding Basic Statistics* by Brase and Brase. Both texts are published by Houghton Mifflin Publishing Company.

In addition, 35 data files from referenced sources are described in the Appendix. These data files are used in class demonstrations in ComputerStat. They are also available as portable MINITAB worksheets and TI-83 files on the DATA DISK.

The Texas Instruments TI-83 Graphics Calculator has many built-in statistical functions and statistical graphs. Data is entered in spreadsheet like format. Data corrections and modifications are convenient since the user can review all data as and after it has been entered. The direct algebra logic of the calculator enables the user to enter arithmetic expressions in the same order as they are written. The TI-83 section of this guide gives procedures with examples for using the statistical features of this calculator and lab activities.

ComputerStat Version 5 is a computer software package specifically written to coordinate with the texts *Understandable Statistics* 5th edition and *Understanding Basic Statistics*. It features Classroom Demonstrations. Users may enter their own data as well. The programs in ComputerStat are interactive and require minimal experience using computers, as well as minimal computer devices. Users simply follow the instructions on the screen and provide the requested information. ComputerStat is very convenient to use. It enables students to explore statistical concepts without the burden of computation.

ComputerStat Version 5 is available in three platforms: a Windows version that works on Windows 3.1 or higher , in a Macintosh® format, and in a MSDOS® format. Adopters of the texts *Understandable Statistics* 6th edition or *Understanding Basic Statistics* qualify for complimentary institutional licenses for ComputerStat. Please contact the publisher Houghton Mifflin for further information.

MINITAB is a professional computer package designed to implement statistical processes. In this guide, we introduce some of the basic menu selections and commands appropriate for exploring statistical concepts presented in the texts *Understandable Statistics* 6th edition and *Understanding Basic Statistics*. Session window commands are given in examples since these commands may be used in almost all platforms of MINITAB. In addition, menu selections for commands may be used in almost all platforms of MINITAB. In addition, menu selections for Windows versions of MINITAB are provided. Graphs using both character graphics and professional graphics are shown. Lab activities are also provided.

The DATA DISK has 35 MINITAB portable worksheets containing real data from referenced sources. These worksheets provide a data from a wide variety of fields such as sports, health, economics, social science, business, and natural science. A description of the data as well as the data themselves are shown in the Appendix of this Guide. The DATA DISK contains the same files in a text format that can be used through the LINK solfware with the TI-83

Acknowledgments

We wish to thank both Texas Instruments Incorporated and Minitab, Inc for cooperation given to us.

C.H.B.
C.P.B.

Information about the Texas Instruments TI-83 graphics calculator can be obtained from

Consumer Relations
Texas Instruments Incorporated
P.O. Box 53
Lubbock, Texas 79408-0053

Phone 1-800-842-2737

Information about the software package ComputerStat Version 5 and the DATA DISK for MINITAB may be obtained from

Houghton Mifflin Publishing Company
222 Berkeley Street
Boston MA 02116
Phone: (617) 351-5000

Information about the software package MINITAB may be obtained from

Minitab, Inc.
3081 Enterprise Drive
State College, PA 16801
USA

Phone: (814) 238-3280
Telex: 881612
Fax: 814-238-4383

Chart to Coordinate *Technology Guide* to *Understanding Basic Statistics*

The contents of *Technology Guide* is organized to coordinate with the text *Understandable Statistics*, 6th edition by Brase and Brase. The order of topics is slightly different for *Understanding Basic Statistics* by Brase and Brase. Use this chart to coordinate *Technology Guide* with *Understanding Basic Statistics.*

Understanding Basic Statistics Chapters	*Technology Guide* Chapters
1 Organizing Data	1 Getting Started 2 Organizing Data
2 Averages and Variation 3 Regression and Correlation	3 Averages and Variation 10 Regression and Correlation Omit testing ρ and Multiple Regression
4 Introduction to Probability 5 The Binomial Probability Distribution	2 Organizing Data: Dice Simulations 4 Elementary Probability Theory 5 The Binomial Probability Distribution
6 Normal Distributions	6 Normal Distributions
7 Introduction to Sampling Distributions	7 Introduction to Sampling Distributions
8 Introduction to Estimation	8 Estimation Omit Confidence Intervals for Mu1 - Mu2 and P1 - P2
9 Hypothesis Testing Involving One Population	9 Hypothesis Testing Include tests of a single mean or proportion
10 Inferences about Differences	8 Estimation Include Difference of Means (Large Sample) and Difference of Proportions 9 Hypothesis Testing Include Tests of Paired Differences, Difference of Means (Large Sample), and Difference of Proportions
11 Additional Topics Using Inference Part I Chi-Square Distribution	11 Chi-Square and F Distribution Omit ANOVA
Part II Inference Relating to Linear Regression	10 Regression and Correlation Include Confidence Intervals for Prediction and Testing ρ.

Contents

Part I TI-83 Graphics Calculator

Part III MINITAB

PART I

TI-83 GRAPHICS CALCULATOR

FOR

UNDERSTANDABLE STATISTICS
SIXTH EDITION

OR

UNDERSTANDING BASIC STATISTICS

CHAPTER 1 GETTING STARTED

ABOUT THE TI-83 GRAPHICS CALCULATOR

All calculators with built in statistical support provide tremendous aid to performing the calculations required to process data for statistical analysis. The Texas Instrument TI-83 Graphics Calculator has many features that are particularly useful in an introductory statistics course. Among the features are

a) Data entry in a spreadsheet like format
The TI-83 has six columns (called lists L_1 L_2 L_3 L_4 L_5 L_6) in which data can be entered. The data in a list can be edited, and new lists can be created by doing arithmetic using old lists.

Sample Data Screen

L1	L2	L3
[illegible]	8.1	1
8.7	6.2	2
6.8	3.6	-1
3.2	.9	2
4.5	1	1
------	------	1
		5

L1(1)=11.2

b) Single variable statistics including: mean, standard deviation, median, maximum, minimum, quartiles 1 and 3, sums

c) Graphs for single variable statistics: histograms, box-and-whisker plots

d) Estimation including confidence intervals using the normal distribution z, or the student's t distribution t for a single mean, and for a difference of means; intervals for a single proportion, and for the difference of proportions

e) Hypothesis testing: single mean (z or t); difference of means (z or t); proportions, difference of proportions, chi-square test of independence; two variances; linear regression; one-way ANOVA

f) Built in probability distributions: normal, student's t, binomial, chi-square, F, poisson, geometric

h) Two variable statistics including linear regression

i) Scatter plots and graph of the least square line

In this *Guide* we will show how to use many features on the TI-83 graphics calculator to aid you as you study particular concepts in statistics. **Lab Activities** coordinated to the text *Understandable Statistics* are also included.

USING THE TI-83

The TI-83 has several functions affiliated with each key. All the items in yellow are accessed by first pressing the yellow [2nd] key. The letters in green are accessed by first pressing the green [ALPHA] key. The four arrow keys [▼] [▲] [►] [◄] enable you to move through a menu screen or along a graph.

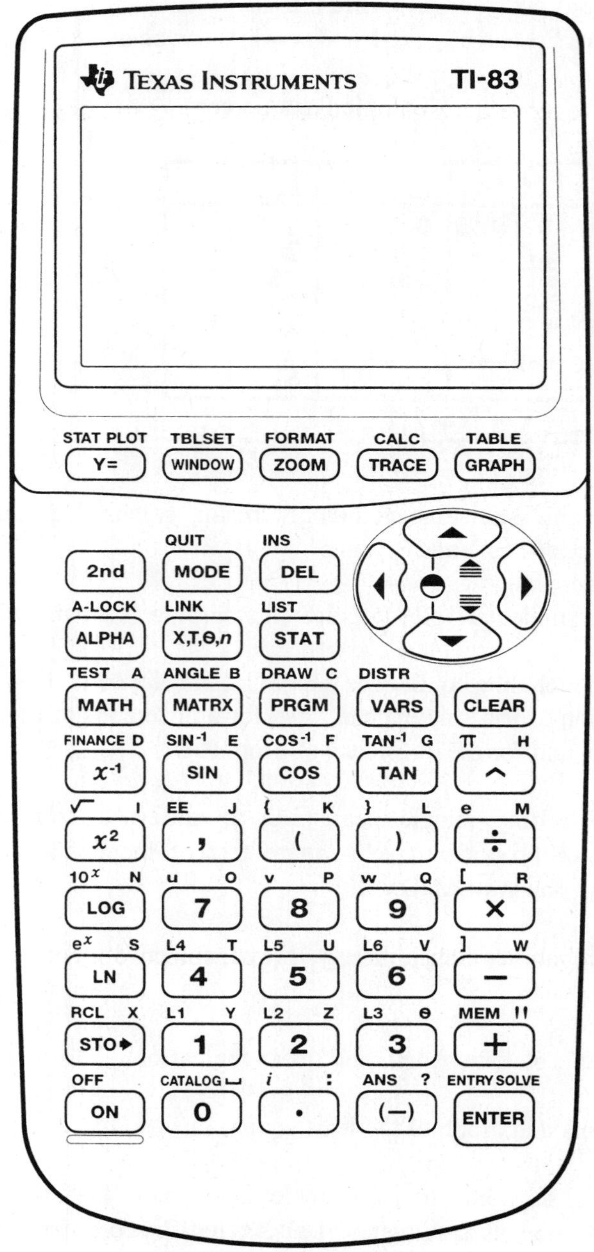

The calculator uses screen menus to access additional operations. For instance, to access the statistics menu, press [STAT]. Your screen should be

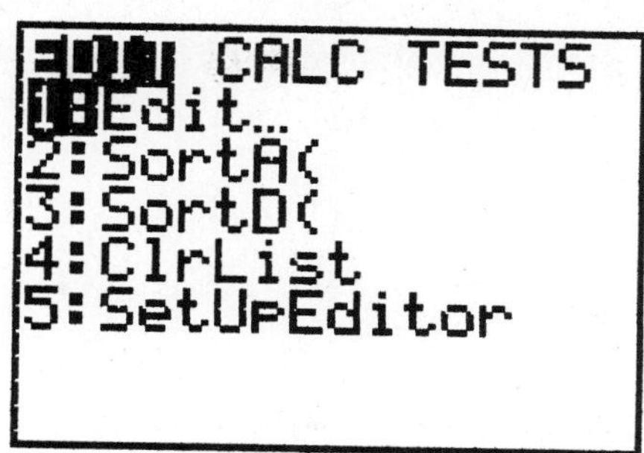

Use the arrow keys to move the highlight to different parts of the screen. Pressing [ENTER] selects a highlighted item.

To leave a screen either press [2nd] [QUIT] or [CLEAR] or select another menu from the keyboard.

Now press [MODE]. When you use your calculator for statistics, the most convenient settings are as shown.

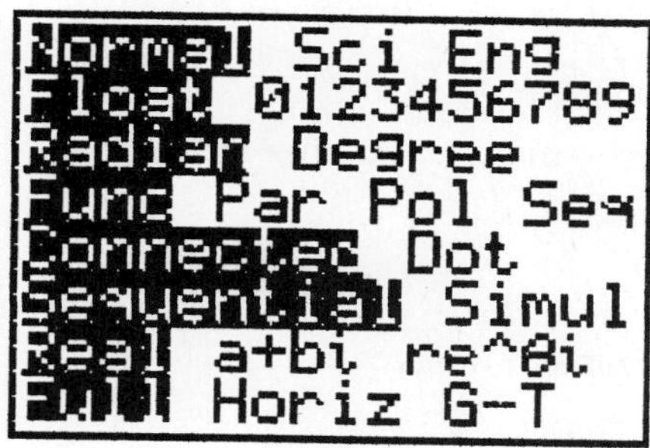

You can change the settings by using the arrow keys to highlight other settings and then pressing [ENTER]

COMPUTATIONS ON THE TI-83

In statistics, you will be evaluating a variety of expressions. The following examples demonstrate some basic keystroke patterns.

Example 1 Evaluate -2(3) + 7

Use the following key strokes

[(-)] [2] [×] [3] [+] [7] [ENTER] The result is 1.

To enter a negative number, be sure to use the key [(-)] rather than the blue subtract key. Notice that the expression -2*3+7 appears on the screen. When you press enter the result 1 is shown on the far right side of the screen.

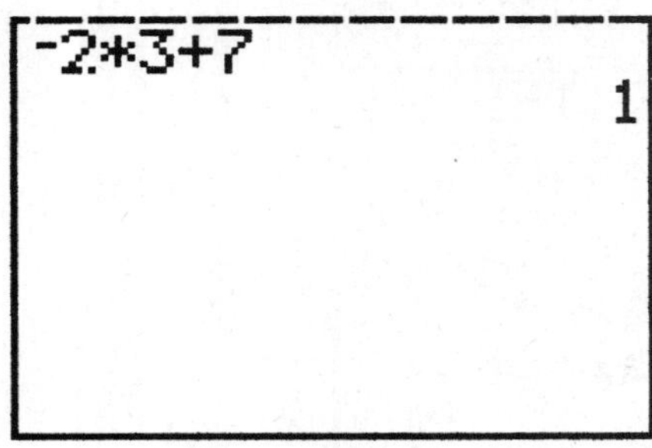

Example 2 Evaluate

a) $\dfrac{10-7}{3.1}$ and round the answer to three places after the decimal.

A reliable approach to evaluating fractions is to enclose the numerator in parentheses, and if the denominator contains more than a single number, enclose it in parentheses as well.

$\dfrac{10-7}{3.1} = (10-7) \div 3.1$ Use the keystrokes

(10 - 7) ÷ 3.1 ENTER Result is .9677419355 which rounds to 0.968

b) $\dfrac{10-7}{\dfrac{3.1}{2}}$ and round the answer to three places after the decimal

Place both numerator and denominator in parentheses (10 - 7) ÷ (3.1 ÷ 2)
Use the keystrokes

The result is 1.935483871 or 1.935 rounded to three places after the decimal

Example 3 Several formulas in statistics require that we take the square root of a value. Note that a left parenthesis (is automatically placed next to the square root symbol. Be sure to close the parentheses after typing in the radicand.

a) Evaluate $\sqrt{10}$ and round the result to three places after the decimal.

2nd √ 10 ENTER The result is 3.16227766 and rounds to 3.162

b) Evaluate $\sqrt{\dfrac{10}{3}}$ and round the result to three places after the decimal.

If the value under the radical contains more than a single number, enclose the expression in parentheses. $\sqrt{}\ (10 \div 3)$

The result rounds to 1.826

Example 4 Some expressions require us to use powers.

a) Evaluate 3.2^2

In this case we are using the second power and we can use the [x^2] key.

The result is 10.24

b) Evaluate 0.4^3

In this case we use the [^] key

The result is 0.064

ENTERING DATA

To use the statistical processes built into the TI-83, we first need to enter data into lists.

Press the [STAT] key. Next we will clear any existing data lists. With EDIT highlighted, arrow down to 4: ClrList and press ENTER.

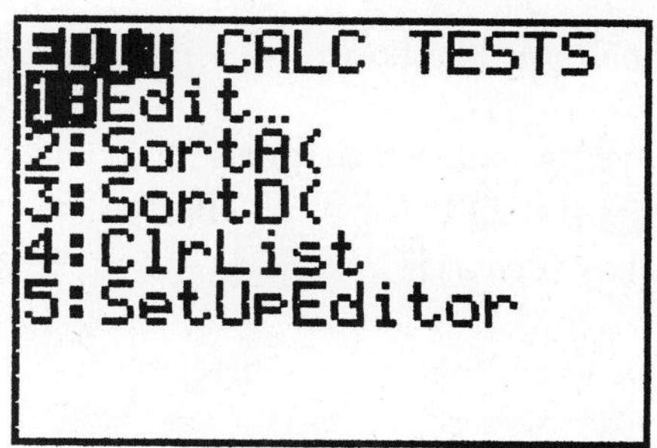

Then type in the six lists separated by commas as shown. Press ENTER

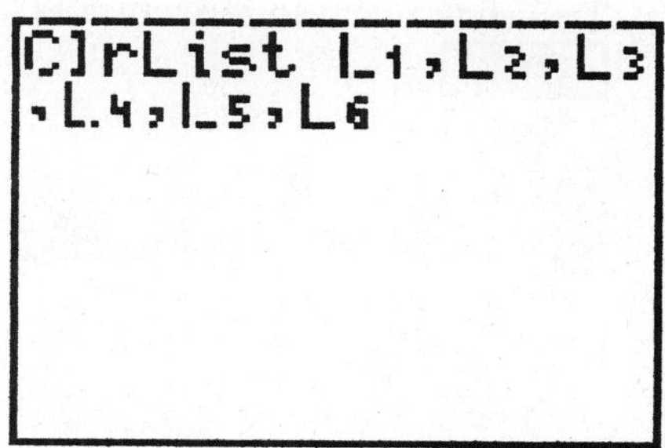

Next press STAT again and select item 1:Edit. You will see the data screen. You are ready to enter data.

Let's enter the numbers

2 5 7 9 11 in list L_1
3 6 9 1 1 in list L_2

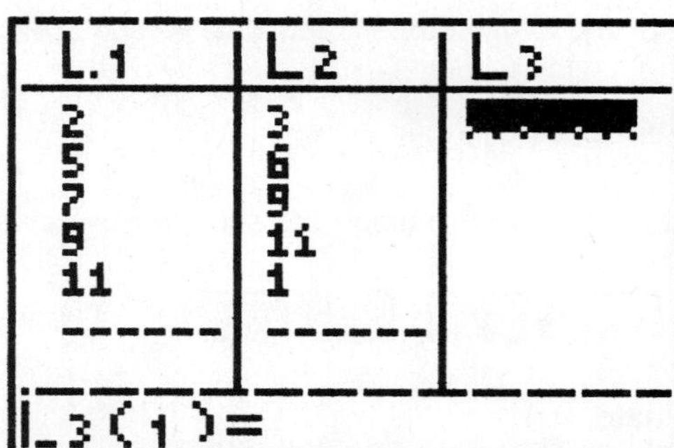

Press ENTER after each number and then when all the numbers are in L_1, use the arrow key to move to L_2.

To correct a data entry error, arrow to the position of the error and enter the correct data value.

We can also create new lists by doing arithmetic on old lists. For instance, let's create data in L_3 by using the formula

$$L_3 = 2L_1 + L_2$$

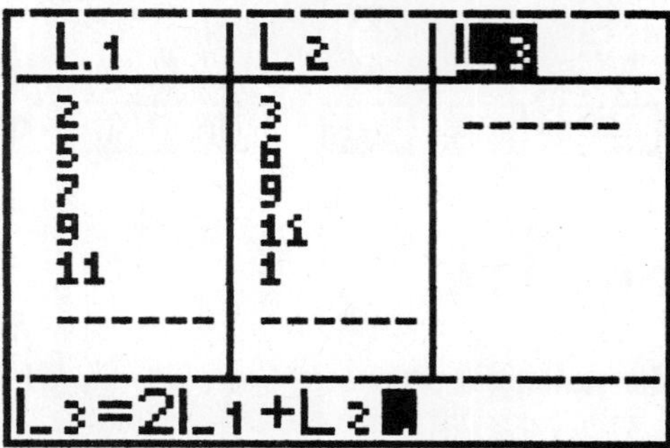

Arrow up to the top and highlight L_3 Then type in $2L_1$ + L_2 and press ENTER

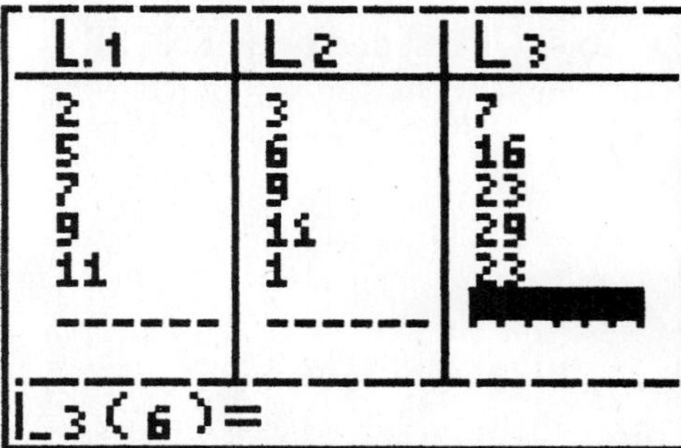

The final result is shown.

To leave the data screen, press 2nd [QUIT] or press another menu key such as STAT

Comments on entering and correcting data.

a) To create a new list from old lists, the lists must all have the same number of entries.

b) To delete a data entry, arrow to the data value you wish to correct and press the DEL key.

c) To insert a data entry into a list, arrow to the position directly below the place you wish to insert the data. Press 2nd [INS] and then enter the data value

Lab Activities to Get Started Using the TI-83

1. Practice doing some calculations using the TI-83. Be sure to use parentheses around a numerator or denominator that contains more than a single number. Round all answers to three places after the decimal.

   a) $\dfrac{5 - 2.3}{1.3}$ Ans. 2.077

   b) $\dfrac{8 - 3.3}{\frac{3}{2}}$ Ans. 3.133

   c) -2(3.4) + 5.8 Ans. -1

   d) -4(-1.7) - 2.1 Ans. 4.7

   e) $\sqrt{5.3}$ Ans. 2.302

   f) $\sqrt{6 + 3(2)}$ Ans. 3.464

   g) $\sqrt{\dfrac{8 - 2.7}{5 - 1}}$ Ans. 1.151

   h) 1.5^2 Ans. 2.25

   i) $(5 - 7.2)^2$ Ans. 4.84

   j) $(0.7)^3(0.3)^2$ Ans. 0.031

2. Enter the following data into the designated list. Be sure to clear all lists first.

   L_1: 3 7 9.2 12 -4

   L_2: -2 9 4.3 16 10

   L_3: $L_1 - 2$

   L_4: $-2L_2 - L_1$

   L_1: Change the data value in the second position to 6

CHAPTER 2 ORGANIZING DATA

RANDOM SAMPLES

(Section 2.1 of *Understandable Statistics*)

The TI-83 graphics calculator has a random number generator. This random number generator can be used in place of a random number table.

It is in the MATH menu under the **PRB** selection. Pressing **ENTER** will place the command on the main computation screen.

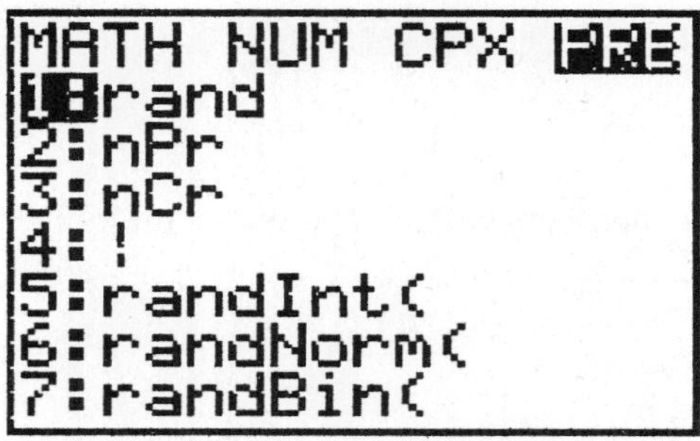

When you press **ENTER** , rand appears on the screen. Press **ENTER** again and a random number between 0 and 1 appears.

To generate a random number that is a whole number with up to 3 digits, multiply the random number by 1000 and take the integer part.

The integer part is found in the MATH menu under **NUM** selection 2:iPart. Press **ENTER** to display the command on the computation screen.

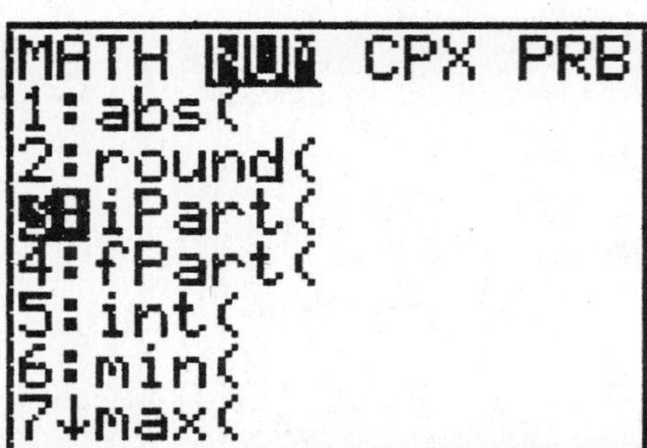

Now let's put all the commands together to generate random numbers that have up to 3 digits.

First display **iPart**. Then open parentheses and type 1000 times **rand** command and close parentheses. Each time you press **ENTER** a new random number will appear. Notice that the numbers might be repeated.

To find random numbers with up to 2 digits, multiply **rand** by 100 instead of by 1000. For random numbers with 1 digit, multiply **rand** by 10.

Example

Use the TI-83 to simulate the outcomes of tossing a die 8 times. Record the results.

In this case, the possible outcomes are the digits 1 through 6. If we generate a random number outside of this range we will ignore it.

Since we want random numbers with 1 digit, we will follow the method prescribed earlier and multiply each random number by 10. Press **ENTER** 6 times.

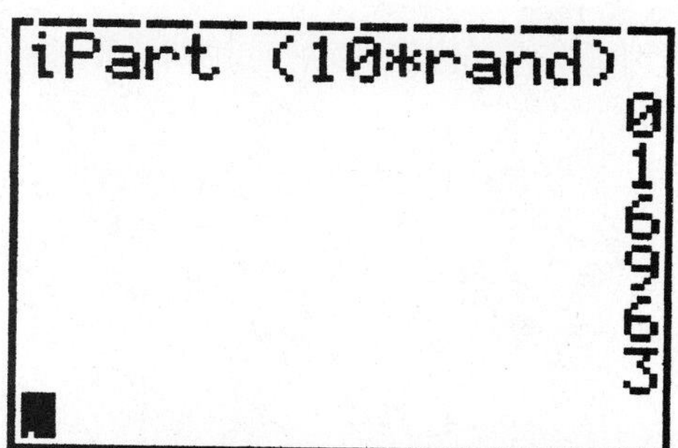

Of the first six values listed, we only use the outcomes 1, 6, 6, 3.

To get the remaining outcomes, keep pressing **ENTER** until two more digits between 1 and 6 appear. When we did it, the next two such digits were 1 and 3. Your results will be different because each time the random number generator is used, it gives a different sequence of numbers.

Lab Activities Using Random Samples

These activates coordinate with Section 2.1 Random Samples of *Understandable Statistics.*

1. Out of a population of 800 eligible county residents, select a random sample of 20 for prospective jury duty. What value should you multiply **rand** by to generate random numbers with 3 digits? List the first 20 numbers corresponding to people for prospective jury duty.

Simulating experiments in which outcomes are equally likely is another important use of random numbers.

2. We can simulate dealing bridge hands by numbering the cards in a bridge deck from 1 to 52. Then we draw a random sample of 13 numbers without replacement from the population of 52 numbers. A bridge deck has 4 suits: Hearts, diamonds, clubs, and spades. Each suit contains 13 cards; those numbered 2 through 10, a jack, a queen, a king, and an ace. Decide how to assign the numbers 1 though 52 to the cards in the deck. Use the random number generator on the TI-83 to get the numbers of the 13 cards in one hand. Translate the numbers to specific cards and tell what cards are in the hand. For a second game the cards would be collected and reshuffled. Using random numbers from the TI-83, determine the hand you might get in a second game.

HISTOGRAMS

The TI-83 graphics calculator draws histograms for data entered in the data lists. The calculator follows the convention that if a data value falls on a class limit, then it is tallied in the frequency of the bar *to the right*. However, if you specify the lower class boundary of the first class and the class width (*see Understandable Statistics* for procedures to find these values), then no data will fall on a class boundary. The following example show you how to draw histograms

Example

Throughout the day from 8 A.M. to 11 P.M. Tiffany counted the number of ads occurring every hour on one commercial T.V. station. The 15 pieces of data are

10	12	8	7	15	6	5	8
8	10	11	13	15	8	9	

To make a histogram with 4 classes:

First enter the data in L_1.

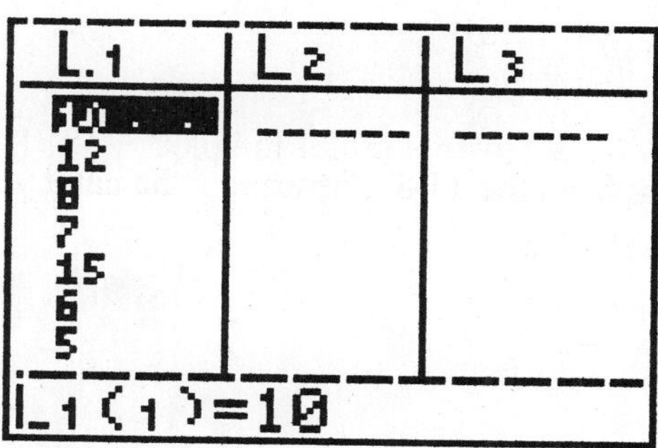

Next we will set the graphing window. However, we need to know the lower class boundary for the first class and the class width. Use techniques found in Section 2.3 of *Understandable Statistics* to find these values.

The smallest data value is 5 so the lower class boundary of the first class is 4.5

The class width is

$$\frac{\text{largest data value} - \text{smallest data value}}{\text{Number of classes}} \quad \text{increased to next integer}$$

$$\frac{15 - 5}{4} = 2.5 \text{ increased to 3}$$

Now we set the graphing window. Press WINDOW key.

Arrow to **Xmin** and use the *lower class boundary* of the first class. For this example, **Xmin = 4.5**

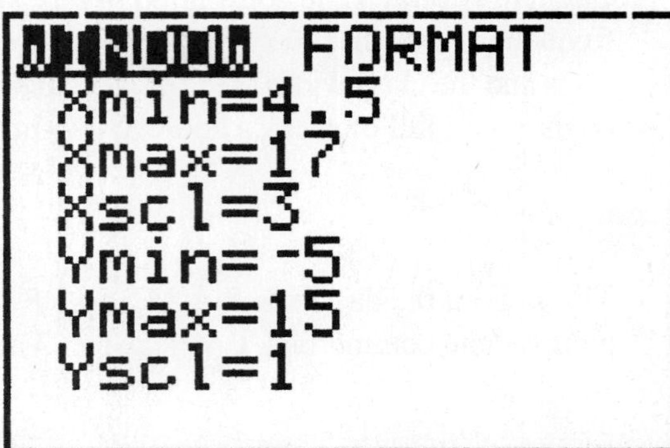

Arrow to **Xscl** and use the *class width*. For this example, **Xscl = 3**

Setting **Xmax = 17** makes sure the histogram fits on the screen since our highest data value was 15.

Setting **Ymin = -5** raises the graph on the screen.

Setting **Ymax = 15** makes sure the histogram fits on the screen.

The next step is to select histogram for the plot. Press 2nd **[STAT PLOT]**

Highlight 1:Plot1 and press **ENTER.**

Then on the next screen highlight **ON** and the **Histogram shape**.

Be sure to highlight the list containing your data. In this example we stored the data in $\mathbf{L_1}$. Each data value occurs only once, so **1** is highlighted in the **FREQ** row.

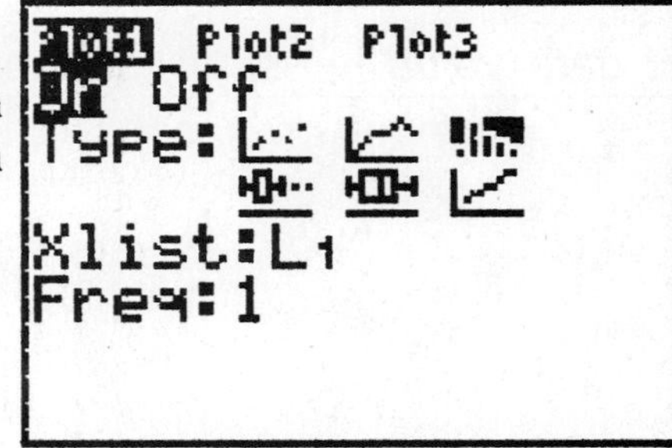

We are ready to graph the histogram. Press GRAPH

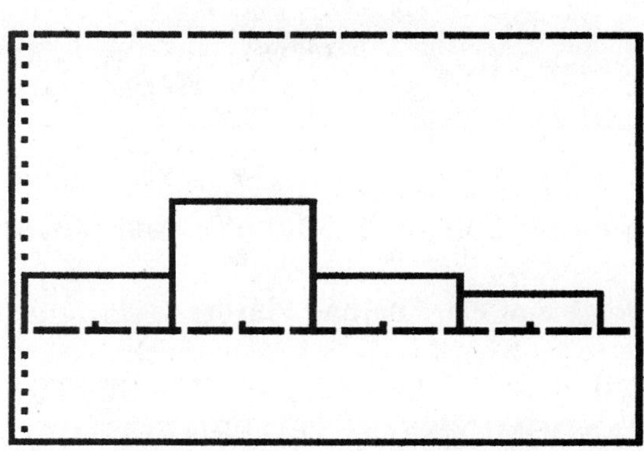

Press TRACE and notice that a blinking cursor appears over the first bar. The class boundaries of the bar are given as the min and max. The frequency is given by n.

Use the right arrow key to move from bar to bar.

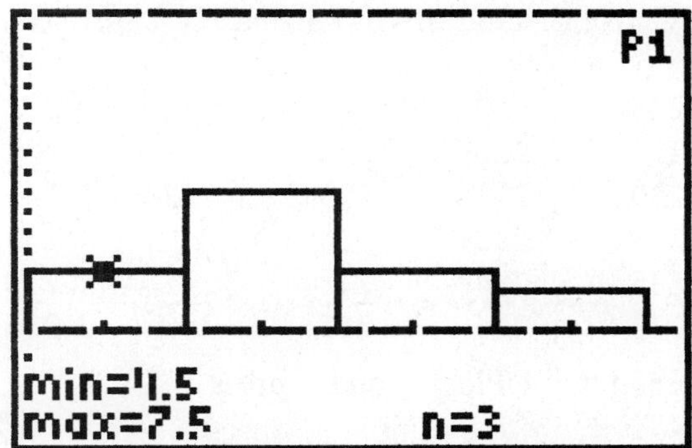

After you have graphed the histogram, you can use the **TRACE** and **arrow** keys to find the class boundaries and frequency of each class. The calculator has done the work of sorting the data and tallying the data for each class.

Lab Activities for Histograms on the TI-83

These activities coordinate with Section 2.3 Histograms and Frequency Distributions of *Understandable Statistics*

1. A random sample of 50 pro football players produced the following data.

Weights of Pro Football Players

THE FOLLOWING DATA REPRESENTS WEIGHTS IN POUNDS OF 50 RANDOMLY SELECTED PRO FOOTBALL LINE BACKERS.

SOURCE: THE SPORTS ENCYCLOPEDIA PRO FOOTBALL 1960-1992

225	230	235	238	232	227	244	222
250	226	242	253	251	225	229	247
239	223	233	222	243	237	230	240
255	230	245	240	235	252	245	231
235	234	248	242	238	240	240	240
235	244	247	250	236	246	243	255
241	245						

Enter this data in L_1. Scan the data for the low and high values. Compute the class width for 5 classes and find the lower class boundary for the first class. Use the TI-83 graphics calculator to make a histogram with 5 classes. Repeat the process for 9 classes. Is the data distribution skewed or symmetric? Is the distribution shape more pronounced with 5 classes or with 9 classes?

2. Explore the other data files found in the **Appendix** of this *Guide* such as.

DISNEY STOCK VOLUME
HEIGHTS OF PRO BASKETBALL PLAYERS
MILES PER GALLON GASOLINE CONSUMPTION
FASTING GLUCOSE BLOOD TESTS
NUMBER OF CHILDREN IN RURAL FAMILIES

Select one of these files and make a histogram with 7 classes. Comment on the histogram shape.

Lab Activities for Histograms on the TI-83 continued

3. a) Consider the data

1	3	7	8	10
6	5	4	2	1
9	3	4	5	2

Place the data in L_1.Use the TI-83 graphics calculator to make a histogram with N = 3 classes. Jot down the class boundaries and frequencies so that you can compare them to part (b)

b) Now add 20 to each data value of part (a). The results are

21	23	27	28	30
26	25	24	22	21
29	23	24	25	22

By using arithmetic in the data list screen, you can create L_2 and $L_2 = L_1 + 20$.

Make a histogram with 3 classes. Compare the class boundaries and frequencies with those obtained in part (a). Are each of the boundary values 20 more than those of part (a)? How do the frequencies compare?

c) Use your discoveries from part (b) to predict the class boundaries and class frequency with 3 classes for the data values below. What do you suppose the histogram will look like?

1001	1003	1007	1008	1010
1006	1005	1004	1002	1001
1009	1003	1004	1005	1002

Would it be safe to say that we simply shift the histogram of part (a) 1000 units to the right?

d) What if we multiply each of the values of part (a) by 10? Will we effectively multiply the entries for class boundaries by 10? To explore this relation create a new list L_3 by the formula $L_3 = 10L_1$. The entries will be

10	30	70	80	100
60	50	40	20	10
90	30	40	50	20

Lab Activities for Histograms on the TI-83 continued

Compare the histogram with three classes to the one from part (a). You will see that there does not seem to be any relation. To see why, look at the class width and compare it to the class width of part (a). The class width is always increased to the next integer value no matter how large the integer data values are. Consequently, the class width for the data in part (d) was increased to 31 instead of to 30.

4. Histograms are not effective displays for some data. Consider the data

1	2	3	6	5	7
9	8	4	12	11	15
14	12	6	2	1	206

Use the TI-83 to make a histogram with 2 classes. Then change to 3 classes, on up to 10 classes. Notice that all the histograms lump the first 14 data into the first class, and the one data value 206 in the last class. What feature of the data causes this phenomenon? Recall that

Class width = (largest data - smallest data)/(number of classes)
increased to the next integer

How many classes would you need before you began to see the first 15 data values distributed among several classes? What would happen if you simply did not include the extreme value 206 in your histogram?

CHAPTER 3 AVERAGES AND VARIATION

1 VARIABLE STATISTICS
(Sections 3.1 and 3.2 of *Understandable Statistics*)

The TI-83 Graphics Calculator provides many of the common descriptive measures for a data set The measures provided with **1-Var Stats** are

Mean $\bar{x}$
Sample standard deviation s_x
Population standard deviation σ_x
Number of data values n
Median
Minimum data value
Maximum data value
Quartile 1
Quartile 3
$\sum x$
$\sum x^2$

Although the mode is not provided directly, the menu choice **SortA(** sorts the data in a specified list from smallest to largest value. By scanning the sorted list, you can find the mode fairly quickly.

Example

At Lazy River College 15 students were selected at random from a group registering the last day of registration. The times (in hours) necessary for these students to complete registration follow:

1.7	2.1	0.8	3.5	1.5	2.6	2.1	2.8
3.1	2.1	1.3	0.5	2.1	1.5	1.9	

Use the TI-83 to find the mean, sample standard deviation, and median. Use the **SortA(** command to sort the data and scan for the mode if it exists.

First enter the data into list L_1

L1	L2	L3
1.7	------	------
2.1		
.8		
3.5		
1.5		
2.6		
2.1		

L1(1)=1.7

Press STAT again, and highlight **CALC** with item **1:1-Var Stats**

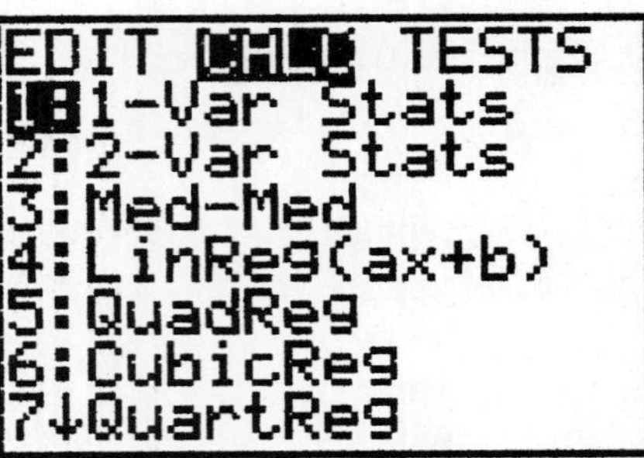

Press **ENTER.** The command **1-Var Stats** will appear on the screen. Type L_1 next to the command to tell the calculator to use the data in list L_1

```
1-Var Stats L1
```

Press **ENTER**. The next two screens contain the statistics for the data in list L_1.

Note the arrow ↓ on the last line. This is an indication that you may use the down arrow key to display more data.

```
1-Var Stats
 x̄=1.973333333
 Σx=29.6
 Σx²=67.68
 Sx=.8136923486
 σx=.7861014919
↓n=15
```

Arrow down. The second full screen shows the rest of the statistics for the data in list L_1

```
1-Var Stats
↑n=15
 minX=.5
 Q1=1.5
 Med=2.1
 Q3=2.6
 maxX=3.5
```


To sort the data, press STAT to return to the edit menu. Highlight **EDIT** and item **2:SortA(** .

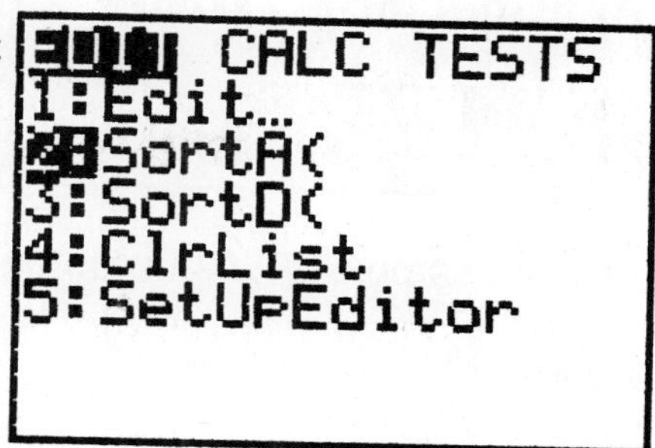

Press **ENTER**. The command **SortA(** appears on the screen. Type in L_1.

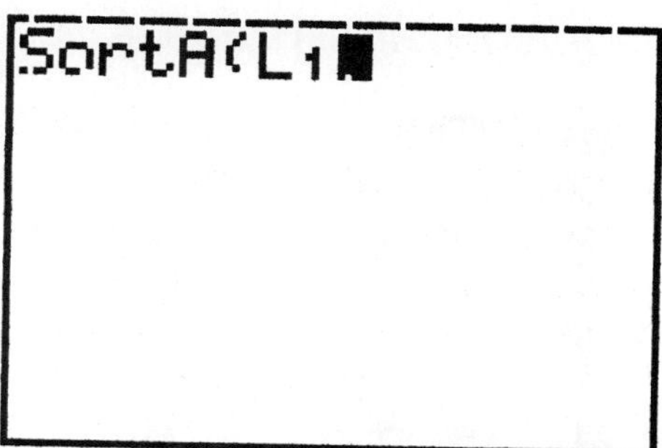

When you press **ENTER** the comment **DONE** appears on the screen.

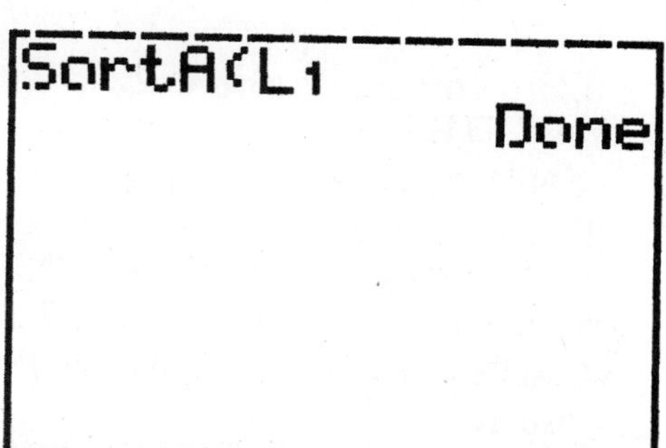

Press STAT and return to the data screen. Notice that the data in list L_1 is now sorted and is in ascending order.

A scan of the data shows that the mode is 2.1 since this data value occurs more than any other.

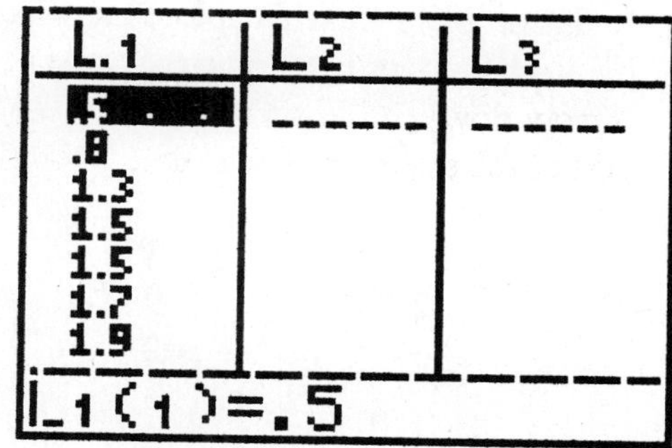

Lab Activities for 1-Variable Statistics on the TI-83

These activities coordinate with Sections 3.1 and 3.2 of *Understandable Statistics*

1. A random sample of 50 pro football players produced the following data.

   Weights of Pro Football Players

   THE FOLLOWING DATA REPRESENTS WEIGHTS IN POUNDS OF 50 RANDOMLY SELECTED PRO FOOTBALL LINE BACKERS.

   SOURCE: THE SPORTS ENCYCLOPEDIA PRO FOOTBALL 1960-1992

225	230	235	238	232	227	244	222
250	226	242	253	251	225	229	247
239	223	233	222	243	237	230	240
255	230	245	240	235	252	245	231
235	234	248	242	238	240	240	240
235	244	247	250	236	246	243	255
241	245						

   Enter this data in L_1. Use **1-Var Stats** to find the mean, median and standard deviation for the weights. Use **SortA** to sort the data and scan for a mode if it exists.

2. Explore some of the other data sets found in the **Appendix** of this *Guide* such as

   DISNEY STOCK VOLUME
   HEIGHTS OF PRO BASKETBALL PLAYERS
   MILES PER GALLON GASOLINE CONSUMPTION
   FASTING GLUCOSE BLOOD TESTS
   NUMBER OF CHILDREN IN RURAL FAMILIES

   Use the TI-83 to find the mean, median, and standard deviation and mode of the data.

3. In this problem we will explore the effects of changing data values by multiplying each data value by a constant, or by adding the same constant to each data value.

   a) Consider the data

1	8	3	5	7
2	10	9	4	6
3	5	2	9	1

   Enter the data into list L_1 and use the TI-83 to find the mode (if it exists), mean,

Lab Activities for 1-Variable Statistics on the TI-83 continued

sample standard deviation, range, and median Make a note of these values since you will compare them to those obtained in parts (b) and (c).

b) Now multiply each data value of part (a) by 10 to obtain the data

10	80	30	50	70
20	100	90	40	60
30	50	20	90	10

Remember you can create these data in L_2 by using the command $L_2 = 10L_1$.

Again useTI-83 to find the mode (if it exists), mean, sample standard deviation, range, and median. Compare these results to the corresponding ones of part (a). Which values changed? Did those that changed change by a factor of 10? Did the range or standard deviation change? Referring to the formulas for these measures (see Section 3.2 of *Understandable Statistics*) can you explain why these values behaved the way they did? Will these results generalize to the situation of multiplying each data entry by 12 instead of by 10? What about multiplying each by 0.5? Predict the corresponding values that would occur if we multiplied the data set of part (a) by 1000.

c) Now suppose we add 30 to each data value of part (a)

32	38	33	35	37
32	40	39	34	36
33	35	32	39	31

To enter this data, create a new list L_3 by using the command $L_3 = L_1 + 30$.

Again use the TI-83 to find the mode (if it exists), mean, sample standard deviation, range, and median. Compare these results to the corresponding ones of part (a). Which values changed? Of those that are different, did each change by being 30 more than the corresponding value of part (a)? Again look at the formulas for range and standard deviation. Can you predict the observed behavior from the formulas? Can you generalize these results? What is we added 50 to each data value of part (a). Predict the values for the mode, mean, median, sample standard deviation and range.

Grouped Data
(Section 3.3 of *Understandable Statistics*)

The TI-83 supports grouped data because it allows you to specify the frequency with which each data value occurs in a separate list. The frequencies must be *whole* numbers. This means that numbers with decimal parts are not permitted as frequencies.

Example

A random sample of 44 automobiles registered in Dallas, Texas show the ages of the cars to be

Age (in years)	Midpoint	Frequency
1 - 3	2	9
4 - 6	5	12
7 - 9	8	20
10-12	11	3

Estimate the mean and standard deviation of the ages of the cars.

Put the midpoints in list L_1 and the corresponding frequencies in list L_2.

L1	L2	L3
2	9	------
5	12	
8	20	
11	3	

L2(5)=

Press STAT and highlight **CALC** This process will take you to the calculations menu shown to the left. Be sure that **CALC** is highlighted and select item **1:1-Var Stats**. Press ENTER

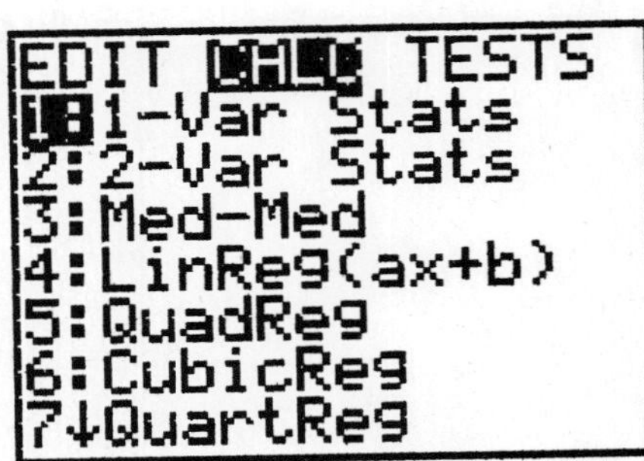

Designate the data lists, with the first list holding the data and the second list holding the frequencies. Press ENTER

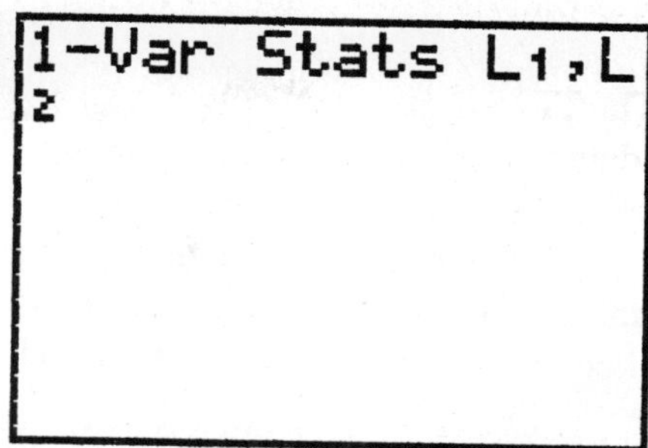

When you press **ENTER** the statistics for the grouped data will appear. Note that n = 44 is the correct value for the number of data.

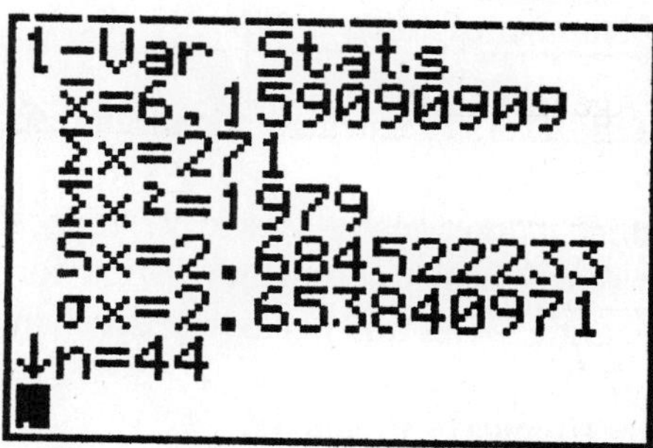

From the last screen, we see the estimate for the mean is 6.16 and the estimate for the sample standard deviation is 2.68.

Lab Activities for Grouped Data on the TI-83

These activities use information from Section 3.3 of *Understandable Statistics*

1. A teacher rating form uses the scale values -2, -1, 0, 1, 2 with -2 meaning strongly disagree and 2 meaning agree strongly. The statement "The professor grades fairly" was answered as follows for all courses in the sociology department.

Score:	-2	-1	0	1	2
Number of responses:	30	125	200	180	72

   Use the TI-83 to calculate the mean response for this statement and the standard deviation.

2. A stock market analyst looked at 80 key stocks and recorded the total gains and losses (in points) per share over a one month period for each stock. The results follow with negative points indicating losses and positive point change indicating gains.

Point Change	Frequency	Point Change	Frequency
-40 to -31	3	0 to 9	20
-30 to -21	2	10 to 19	14
-20 to -11	10	20 to 29	7
-10 to -1	19	30 to 39	5

   First find the class midpoints. Then use the TI-83 to estimate the mean point change and standard deviation for this *population* of stocks.

Box-and-Whisker Plot
(Section 3.4 of *Understandable Statistics*)

A box-and-whisker plot is a display from Exploratory Data Techniques that is supported by the TI-83. The box-and-whisker plot is based on the five number summary values found under **1-Var Stats**:

Low value
Quartile 1, Q_1
Median
Quartile 2, Q_2
High value

The box-and-whisker plot is found in the **STAT PLOT** menu.

Example

Let's make a box-and-whisker plot using the data about time it takes to register for Lazy River College students who waited till the last day to register. These data (in hours) are

1.7	2.1	0.8	3.5	1.5	2.6	2.1	2.8
3.1	2.1	1.3	0.5	2.1	1.5	1.9	

In the previous example we found the **1-Var Stats** for this data.

First enter the data into list L_1

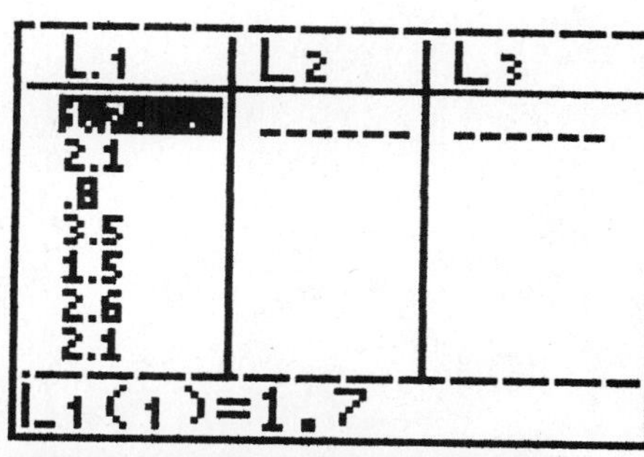

Press WINDOW to set the graphing window.

Use Xmin = 0.5 since that is the smallest data.
Use Xmax = 3.5 since that is the largest data.
Use Xscl = 1
Use Ymin = -5 to position the graph in the window.
Use Ymax =10 to position the graph in the window.

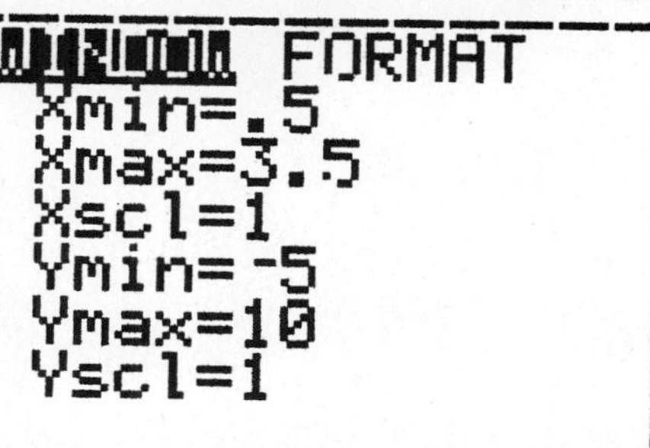

Press 2nd [STAT PLOT] and highlight **1:Plot**. Press **ENTER**

Highlight **ON**, and select the **box plot**. The Xlist should be L_1 and Freq should be **1**.

Press GRAPH.

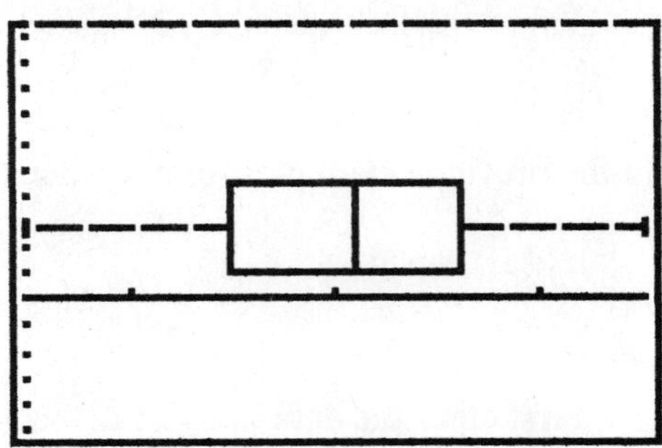

Press TRACE and use the **arrow** keys to display the five number summary.

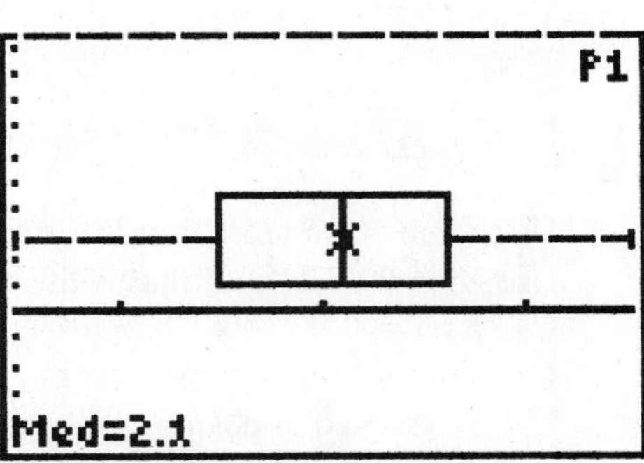

Lab Activities for Box-and-Whisker Plot

These activities coordinate with Sections 2.3, 3.1, 3.3 and 3.4 of *Understandable Statistics*

You now have many descriptive tools available: Histograms, box-and-whisker plots, averages, and measures of variation such as the standard deviation. When you use all of these tools, you get a lot of information about the data.

1. One of the data sets included in the **Appendix** gives the miles per gallon gasoline consumption for a random sample of 55 makes and models of passenger cars.

Miles per Gallon Gasoline Consumption

THE FOLLOWING DATA REPRESENTS MILES PER GALLON GASOLINE CONSUMPTION (HIGHWAY) FOR A RANDOM SAMPLE OF 55 MAKES AND MODELS OF PASSENGER CARS.

SOURCE: ENVIRONMENTAL PROTECTION AGENCY

30	27	22	25	24	25	24	15
35	35	33	52	49	10	27	18
20	23	24	25	30	24	24	24
18	20	25	27	24	32	29	27
24	27	26	25	24	28	33	30
13	13	21	28	37	35	32	33
29	31	28	28	25	29	31	

Enter this data into list L_1, and then use the TI-83 to make a histogram, a box-and-whisker plot, compute the mean, median, and sample standard deviation. Based on the information you obtain respond to the following questions:

a) Is the distribution skewed or symmetric? How is this shown in both the histogram and the box-and-whisker plot?
b) Look at the box-and-whisker plot. Are the data more spread out above the median or below the median?
c) Look at the histogram and estimate the location of the mean on the horizontal axis. Are the data more spread out above the mean or below the mean?
d) Do there seem to be any data values that are unusually high or unusually low? If so, how do these show up on a histogram or on a box-and-whisker plot?
e) Pretend that you are writing a brief article for a newspaper. Describe the information about the data in the class demonstration you selected in non technical terms. Be sure to make some comments about the "average" of the data measurements and some comments about the spread of the data.

Lab Activities for Box-and-Whisker Plot continued

2. a) Consider the test scores of 30 students in a political science class.

85	73	43	86	73	59	73	84	62	100
75	87	70	84	97	62	76	89	90	83
70	65	77	90	84	80	68	91	67	79

For this population of test scores find the mode, median, mean, range, variance, standard deviation, CV, and the five number summary and make a box-and-whisker plot. Be sure to record all of these values so you can compare them to the results of part (b).

b) Suppose Greg was in the political science class of part (a). Suppose he missed a number of classes because of illness, but took the exam anyway and made a score of 30 instead of 85 as listed as the first entry of the data in part (a). Again, use the program to find the mode, median, mean, range, variance, standard deviation, CV, and the five number summary and make a box-and-whisker plot using the new data set. Compare these results to the corresponding results of part (a). Which average was most affected: mode, median, or mean? What about the range, standard deviation, and coefficient of variation? How do the box-and-whisker plots compare?

c) Write a brief essay in which you use the results of parts (a) and (b) to predict how an extreme data value affects a data distribution. What do you predict for the results if Greg's test score had been 80 instead of 30 or 85?

Chapter 4 Elementary Probability Theory

There are no specific TI-83 activities for the basic rules of probability.

However, notice that the TI-83 has menu items for factorial notation, combinations $C_{n,r}$ and permutations $P_{n,r}$.

To find these functions, press the MATH key and highlight PRB.

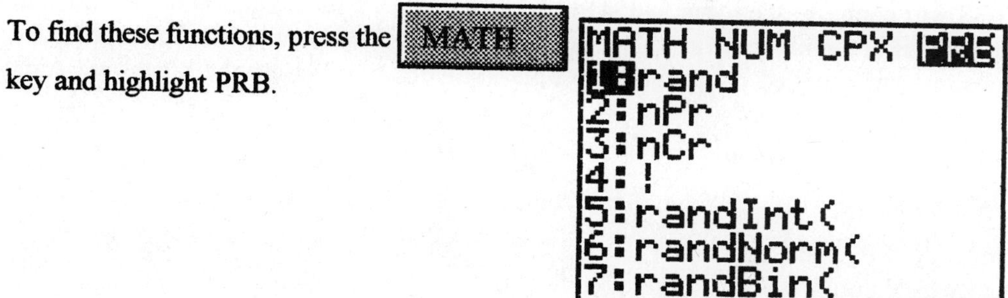

To compute $C_{12,6}$ first clear the screen and type 12. Then use the MATH key and select PRB and item **3:nCr**. Type the number 6 and then press ENTER

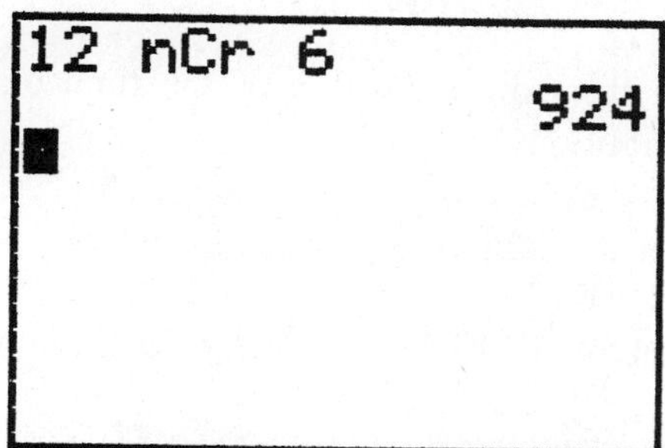

The menu option **nPr** works in a similar fashion and gives you the value of $P_{n,r}$

CHAPTER 5 THE BINOMIAL PROBABILITY DISTRIBUTION AND RELATED TOPICS

DISCRETE PROBABILITY DISTRIBUTIONS

(Section 5.1 of *Understandable Statistics*)

The TI-83 will compute the mean and standard deviation of a discrete probability distribution. Just enter the values of the random variable x in list L_1 and the corresponding probabilities in list L_2. Use the techniques of grouped data and 1-variable statistics to compute the mean and standard deviation.

Example

How long do we spend on hold when we call a store? One study produced the following probability distribution with the times recorded to the nearest minute.

X, Time on hold:	0	1	2	3	4	5
P(X):	0 .15	0.25	0.40	0.10	0.08	0.02

Find the expected value of the time on hold and the standard deviation of the probability distribution.

Enter the data times in L_1 and enter the probabilities in list L_2.

Press the STAT key, choose CALC and use 1-var statistics with list L_1 and L_2.

The results are displayed. The expected value is $\overline{x} = 1.70$ minutes with standard deviation $\sigma = 0.934$.

```
1-Var Stats
 x̄=1.708333333
 Σx=369
 Σx²=819
 Sx=.9366565974
 σx=.9344858955
↓n=216
```

Lab Activities for Discrete Probability Distributions

1. The probability distribution for scores on a mechanical aptitude test is

Score X:	0	10	20	30	40	50
P(X):	.130	. 200	. 300	.170	. 120	.080

Use the TI-83 to find the expected value and standard deviation of the probability distributions.

2. Hold Insurance has calculated the following probabilities for claims on a new $20,000 vehicle for one year if the vehicle is driven by a single male under 25.

# Claim, X:	0	1000	5000	100000	200000
P(X):	0.63	0.24	0.10	0.02	0.01

is the expected value of the claim in one year. What is the standard deviation of the claim distribution? What should the annual premium be to include $400 in overhead an profit as well as the expected claim?

USING THE FORMULA TO COMPUTE PROBABILITIES
(Section 5.2 *of Understandable Statistics*)

For a binomial distribution with

n trials; r successes; probability of success on a single trial p

probability of failure on a single trial q = 1 - p

The formula for the probability of r successes out of n trials is

$$P(r) = C_{n,r}\, p^r\, q^{n-r}$$

To compute probabilities for specific values of n, r, and p, we can use the TI-83 graphics calculator.

Example

Consider a binomial experiment with 10 trials and probability of success on a single trial p = 0.72. Compute the probability of 7 successes out of n trials.

On the TI-83, the combinations function **nCr** is found in the MATH menu under **PRB**.

MATH NUM HYP PRB
1:rand
2:nPr
3:nCr
4:!

To use the function, we first type the value of n and then enter the **nCr** function from the **MATH** menu. Next we type the value of r.

For the example, n = 10 and r = 7, p = 0.72 and q = 0.28. By the formula, we raise .72 to the power 7 and .28 to the power 3.

The probability of 7 successes is 0.264

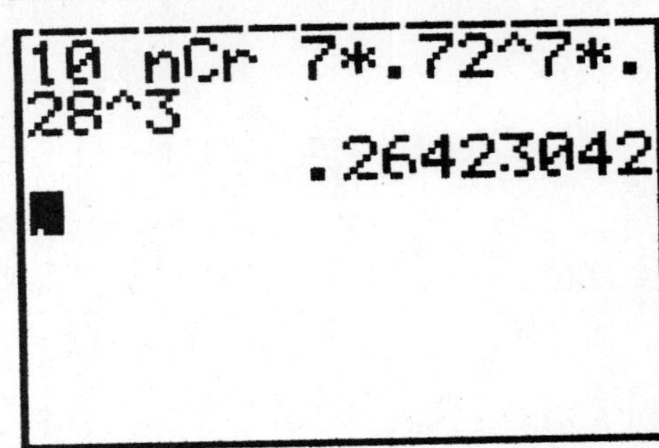

Using the Probability Distributions on the TI-83

The TI-83 fully supports the binomial distribution and has it built in as a function.

To access the DISTR menu that contains the probability functions press 2nd function and the [VARS] key. Then select item **0:binomial pdf(**. and press ENTER.

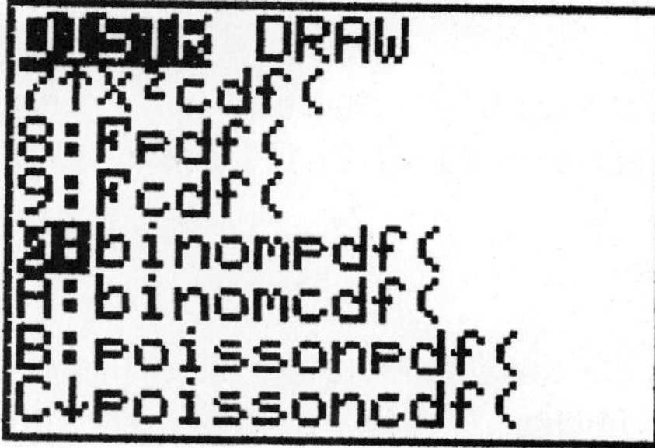

Type in the number of trials, followed by the probability of success on a single trial, followed by the number of successes. Separate each entry by a comma. Press enter.

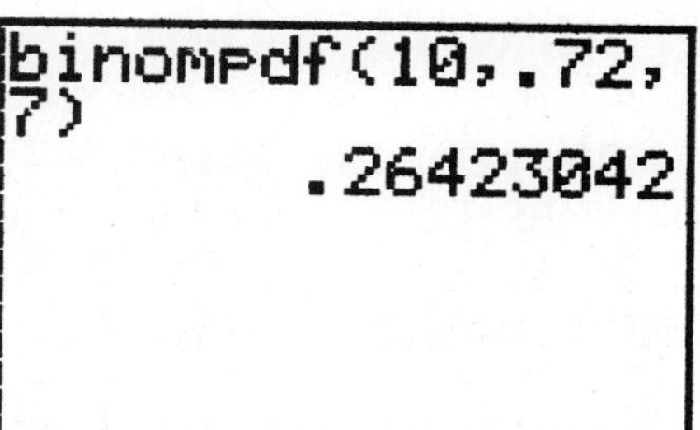

Lab Activities for Binomial Probability Distribution

Consider a binomial distribution with n = 8 and p = 0.43. Use the formula to find the probability of r successes for r from 0 through 8. Use the built in binomial distribution to do the same activity.

CHAPTER 6 NORMAL DISTRIBUTIONS

CONTROL CHARTS

(Section 6.1 of *Understandable Statistics*)

Although the TI-83 does not have a control chart option built into **STAT PLOT** we can use one the features in **STAT PLOT** combined with the regular graphing options to create a control chart.

Example

Consider the data from the data files in the Appendix regarding yield of wheat at Rothamsted Experimental Station over a period of thirty consecutive years. Use the TI-83 to make a control chart for this data using the target mean and standard deviation values.

YIELD OF WHEAT AT ROTHAMSTED EXPERIMENT STATION, ENGLAND

THE FOLLOWING DATA REPRESENT ANNUAL YIELD OF WHEAT IN TONNES (ONE TON = 1.016 TONNE) FOR AN EXPERIMENTAL PLOT OF LAND AT ROTHAMSTED EXPERIMENT STATION U.K. OVER A PERIOD OF THIRTY CONSECUTIVE YEARS.

SOURCE: ROTHAMSTED EXPERIMENT STATION U.K.

WE WILL USE THE FOLLOWING TARGET PRODUCTION VALUES:
TARGET MU = 2.6 TONNES
TARGET SIGMA = 0.40 TONNES

1.73	1.66	1.36	1.19	2.66	2.14	2.25	2.25	2.36	2.82
2.61	2.51	2.61	2.75	3.49	3.22	2.37	2.52	3.43	3.47
3.20	2.72	3.02	3.03	2.36	2.83	2.76	2.07	1.63	3.02

First we enter the data by row. In list L_1 put the year numbers 1 through 30. In list L_2 put the corresponding annual yield. Again, read the data by row.

L1	L2	L3
1	1.73	------
2	1.66	
3	1.36	
4	1.19	
5	2.66	
6	2.14	
7	2.25	

L1(1)=1

Then press 2nd [STAT PLOT] and select **1:Plot1**

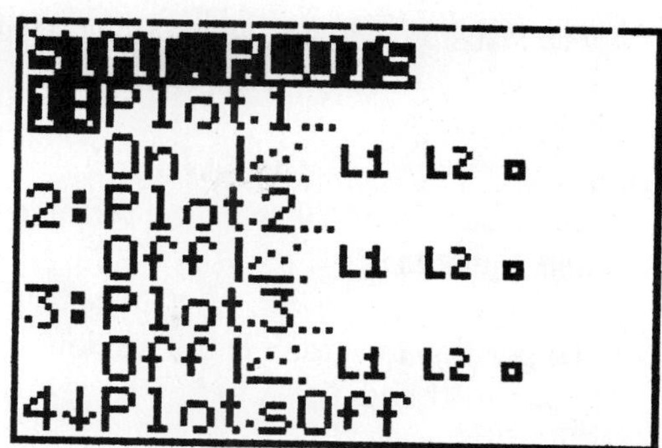

Highlight **ON**. Select scatter plot for type; L_1 for Xlist and L_2 for Ylist. Select the symbol you like for Mark.

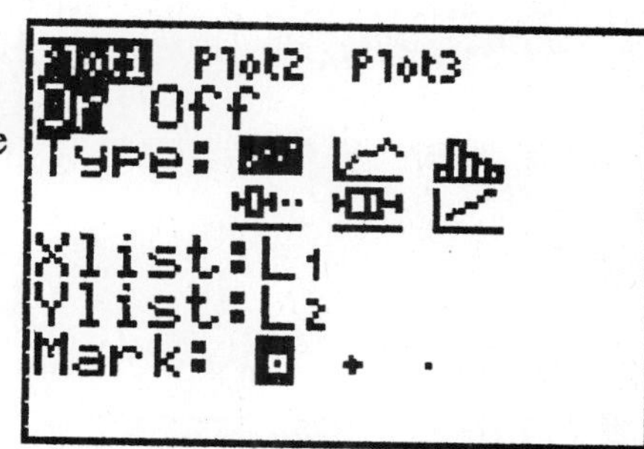

Next pres Y=.

For Y1 enter the target mean 2.6

For Y2 enter the mean + 2 standard deviations. In this case, enter 2.6 + 2*.4

For Y3 enter the mean - 2 standard deviations. In this case, enter 2.6 - 2*.4

For Y4 enter the mean + 3 standard deviations. In this case, enter 2.6 + 3*.4

For Y5 enter the mean - 3 standard deviations. In this case enter 2.6 - 3*.4

```
Y1=2.6
Y2=2.6+2*.4
Y3=2.6-2*.4
Y4=2.6+3*.4
Y5=2.6-3*.4
Y6=
Y7=
Y8=
```

The last step is to set the graphing window. Press WINDOW.

Since there were 30 years, set Xmin to 1 and Xmax to 30

To position the graph in the window, set Ymin to -1 and Ymax to 5

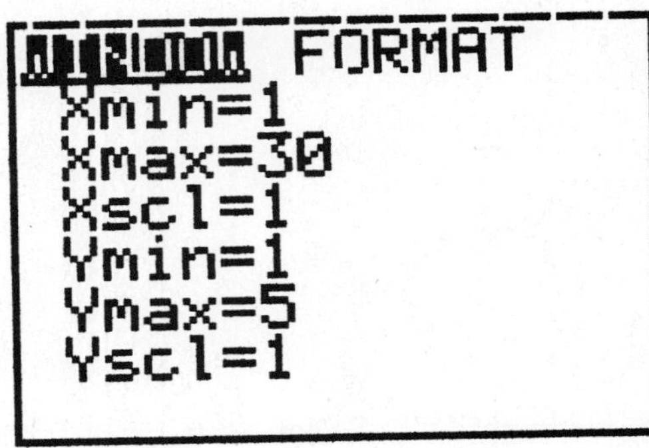

When your press ENTER the control chart appears.

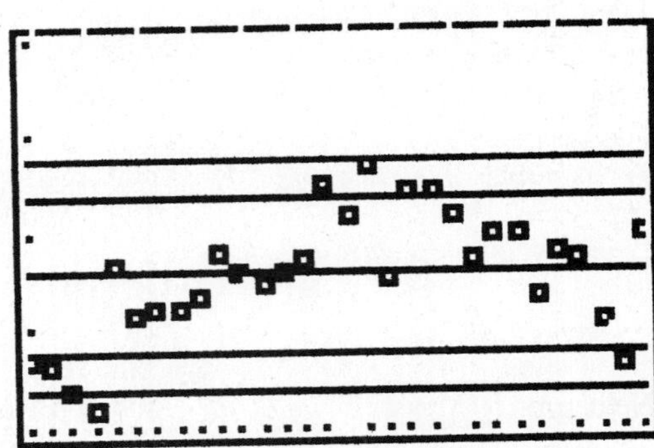

Lab Activities for Control Charts

These activities coordinate with Section 6.1 of *Understandable Statistics*.

1. Look in the Appendix and find the data file for **Futures quotes for the Price of Coffee Beans**. Make a control chart using the data and the target mean and target standard deviation given. Read the data by row. Are there any out of control signals? Explain.

2. Look in the Appendix and find the data file for **Incidence of Melanoma Tumors**. Make a control chart using the data and the target mean and target standard deviation given. Read the data by row. Are there any out of control signals? Explain.

Finding the Area Under Any Normal Curve
(Section 6.3 of *Understandable Statistics*)

The TI-83gives areas under any normal distribution and shades the specified area.

Press 2nd [VARS] to access the probability distributions. Select **2:normalcdf(** and press enter.

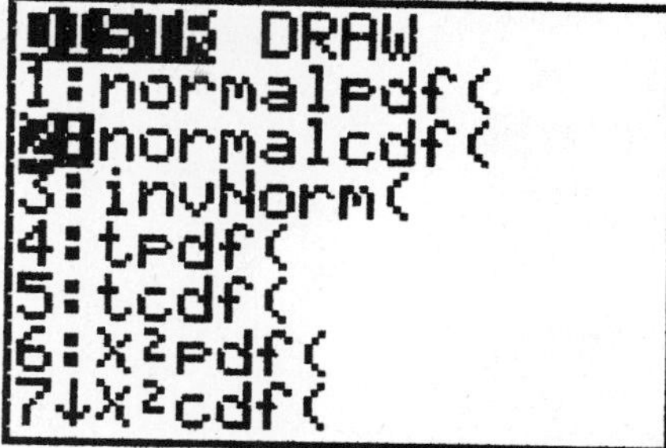

Then type in the lower bound, upper bound, μ, σ in that order separated by commas. To find the area under the normal curve with μ = 10 and σ = 2 between 4 and 12, select normalcdf(and enter the values as shown. Then press ENTER.

```
normalcdf(4,12,1
0,2)
           .8399947732
```

Drawing the normal distribution

To draw the graph, first set the WINDOW to accommodate the graph. Press the WINDOW button and enter values as shown.

```
WINDOW
 Xmin=2
 Xmax=18
 Xscl=3
 Ymin=-.1
 Ymax=.3
 Yscl=1
 Xres=1
```

Press DISTR again and highlight DRAW. Select **1:ShadeNorm(**

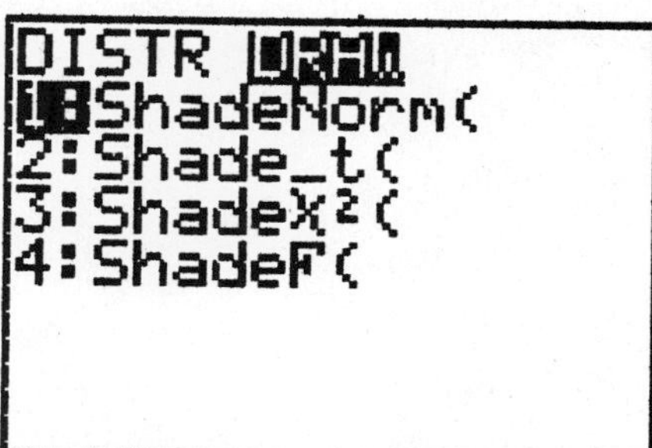

Again enter the lower limit, upper limit, μ, σ separated by commas.

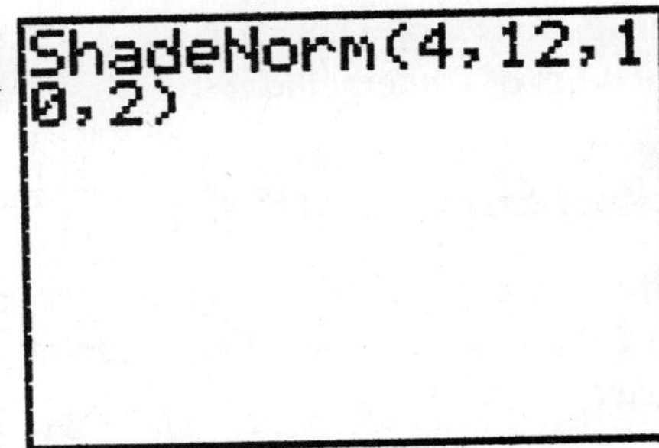

Finally press ENTER.

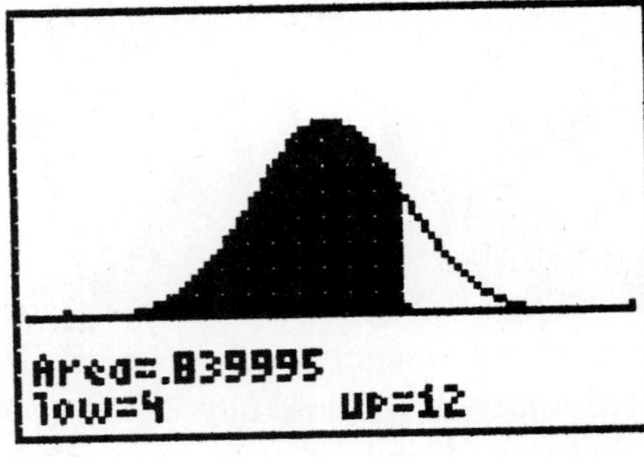

Lab Activities for Areas Under Any Normal Curve

1. Find the area under a normal curve with mean 10 and standard deviation 2 between 7 and 9. Show the shaded region.

2. To find the area in the right tail of a normal distribution, select the value of 5σ for the upper limit of the region. Find the area under a normal curve with mean 10 and standard deviation 2 that lies to the right of 12.

3. Consider a random variable x that follows a normal distribution with mean 100 and standard deviation 15. Shade the regions corresponding to the probabilities and find
   (a) $P(x < 90)$
   (b) $P(70 < x < 100)$
   (c) $P(x > 115)$
   (c) If the random variable was larger than 145, would that be an unusual event? Explain by computing $P(x > 145)$ and commenting on the meaning of the result.

CHAPTER 7 INTRODUCTION TO SAMPLING DISTRIBUTIONS

In this chapter use the TI-83 graphics calculator to do computations. For example, to compute the z score corresponding to a raw score from an $\bar{x}$ distribution, we use the formula

$$z = \frac{\bar{x} - \mu}{\frac{\sigma}{\sqrt{n}}}$$

To evaluate z, be sure to use parentheses as necessary.

Example

If a random sample of size 40 is taken from a distribution with mean $\mu = 10$ and standard deviation $\sigma = 2$, find the z score corresponding to $\bar{x} = 9$. We use the formula

$$z = \frac{9 - 10}{\frac{2}{\sqrt{40}}}$$

Key in the expression using parentheses

(9 - 10) ÷ (2 ÷ √ 40) enter

The result rounds to z = -3.16

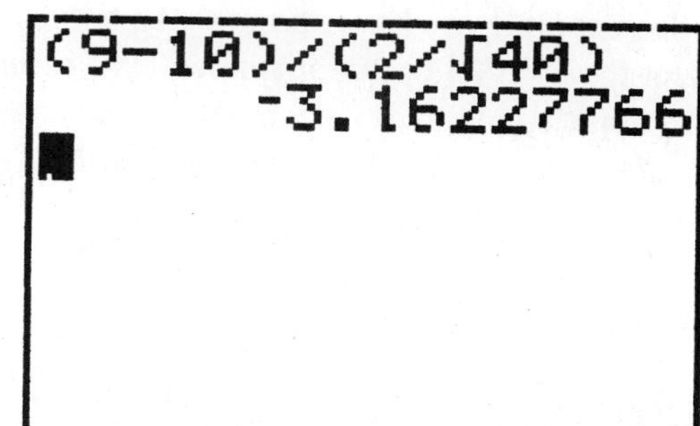

CHAPTER 8 ESTIMATION

CONFIDENCE INTERVALS FOR A POPULATION MEAN MU
(Sections 8.1 and 8.2 of *Understandable Statistics*)

The TI-83 fully supports confidence intervals. To access the confidence interval choices, press the STAT key and select TESTS. The confidence interval choices are found in items 7 through B.

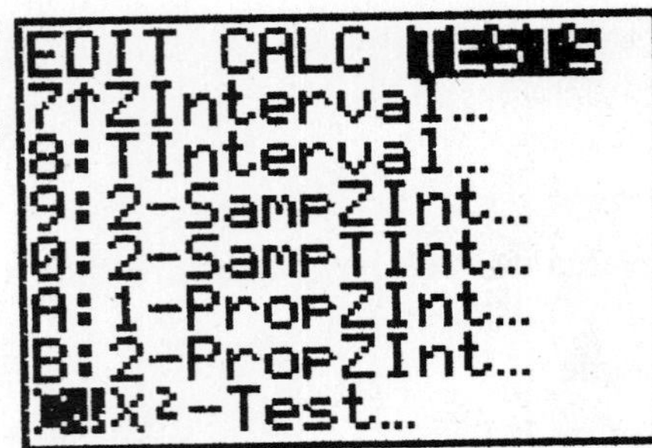

Example (Large Sample): Suppose a random sample of 250 credit card bills showed an average balance of $1200 with standard deviation of $350. Find a 95% confidence interval for the population mean credit card balance.

Since we have a large sample, we will use the normal distribution. Select item **7:ZInterval...**
In this example, we have summary statistics, so we will select the STATS option for input. We estimate σ by the sample standard deviation 350. Then enter the value of $\bar{x}$ and the sample size n. Use 95 for the C-Level.

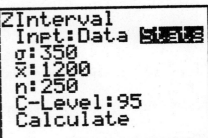

```
ZInterval
 Inpt:Data Stats
 σ:350
 x̄:1200
 n:250
 C-Level:95
 Calculate
```

Highlight Calculate and press Enter to get the results. Notice that the interval is given using standard mathematical notation for an interval. The interval for μ goes from $1156.6 to $1232.4

```
ZInterval
 (1156.6,1243.4)
 x̄=1200
 n=250
```

Example (Small Sample): A random sample of 16 wolf dens showed the number of pups in each to be

5 8 7 5 3 4 3 9
5 8 5 6 5 6 4 7

Find a 90% confidence interval for the population mean number of pups in such dens.

In this case we have raw data, so enter the data in list L_1 using the EDIT option of the STAT key. Since we have a small sample, we use the t distribution. Select item **8:TInterval**.

Since we have raw data, select the DATA option for Input The data is in list L_1, and occurs with frequency 1. Enter 90 for the C-Level.

```
TInterval
 Inpt:Data Stats
 List:L1
 Freq:1
 C-Level:90
 Calculate
```

Highlight Calculate and press ENTER. The result is the interval from 4.84 pups to 6.41 pups.

```
TInterval
 (4.8431,6.4069)
 x̄=5.625
 Sx=1.784189825
 n=16
```

Lab Activities for CONFIDENCE INTERVALS FOR A POPULATION MEAN MU

These activities coordinate with Sections 8.1 and 8.2 of *Understandable Statistics*.

1. Market Survey was hired to do a study for a new soft drink, Refresh. A random sample of 20 people were given a can of Refresh and asked to rate it for taste on a scale of 1 to 10 (with 10 being the highest rating). The ratings were

   5 8 3 7 5 9 10 6 6 2
   9 2 1 8 10 2 5 1 4 7

   Find an 85% confidence interval for the population mean rating of Refresh.

2. Suppose a random sample of 50 basketball players showed the average height to be 78 inches with sample standard deviation 1.5 inches.

   a) Find a 99% confidence interval for the population mean height.
   b) Find a 95% confidence interval for the population mean height.
   c) Find a 90% confidence interval for the population mean height.
   d) Find an 85% confidence interval for the population mean height.

e) What do you notice about the length of the confidence interval as the confidence level goes down? If you used a confidence level of 80%, do you expect the confidence interval to be longer or shorter than that of 85%? Run the program again to verify your answer.

CONFIDENCE INTERVALS FOR THE PROBABILITY OF SUCCESS P IN A BINOMIAL DISTRIBUTION

(Section 8.3 of *Understandable Statistics*)

To find a confidence of a proportion, press the STAT key and use option **A:1-PropZInt...** under TESTS. Notice that the normal distribution will be used.

Example: The public television station BPBS wants to find the percent of its viewing population who give donations to the station. A random sample of 300 viewers were surveyed and it was found that 123 made contributions to the station. Find a 95% confidence interval for the probability that a viewer of BPBS selected at random contributes to the station.

The letter x is used to count the number of successes (the letter r is used in the text). Enter 123 for x and 300 for n. Use 95 for the C-Level.

```
1-PropZInt
 x:123
 n:300
 C-Level:95
 Calculate
```

Highlight Calculate and press ENTER. The result is the interval from 0.35 to 0.46

```
1-PropZInt
 (.35434,.46566)
 p̂=.41
 n=300
```

Lab Activities for CONFIDENCE INTERVALS FOR THE PROBABILITY OF SUCCESS P IN A BINOMIAL DISTRIBUTION

These activities coordinate with Section 8.3 of *Understandable Statistics*.

1. There are many types of errors that will cause a computer program to terminate or give incorrect results. One type of error is punctuation. For instance, if a comma is inserted in the wrong place, the program might not run. A study of programs written by students in a beginning programming course showed that 75 out of 300 errors selected at random were punctuation errors. Find a 99% confidence interval for the proportion of errors made by beginning programming students that are punctuation errors. Next find a 90% confidence interval. Is this interval longer or shorter?

2. Sam decided to do a statistics project to determine a 90% confidence interval for the probability that a student at West Plains College eats lunch in the school cafeteria. He surveyed a random sample of 12 students and found that 9 ate lunch in the cafeteria. Can Sam use the program to find a confidence interval for the population proportion of students eating in the cafeteria? Why or why not? Try the program with N = 12 and R = 9. What happens? What should Sam do to complete his project?

CONFIDENCE INTERVALS FOR MU1-MU2 (INDEPENDENT SAMPLES) CONFIDENCE INTERVALS FOR P1-P2 (LARGE SAMPLES)

(Section 8.5 of *Understandable Statistics*)

For tests of difference of means press the STAT key, highlight TESTS and use the option **9:2-Sample ZInt** for confidence intervals for the difference of means large sample. Use option **0:2-SampTInt** for confidence intervals for the difference of means small sample. Option **B:2-PropZInt** gives confidence intervals for the difference of two proportions.

For the difference of means, you have the option of using summary statistics or raw data that is in lists. Under Inpt:, Data allows you to use raw data from a list, while Stats lets you use summary statistics.

When finding confidence intervals for difference of means, large samples or small samples, use sample estimates s_1 and s_2 for corresponding population standard deviations σ_1 and σ_2. For confidence intervals for difference of means using small samples, be sure to choose YES for Pooled.

Example (Difference of means large samples) A random sample of 45 pro football players showed a sample mean height of 6.18 feet with standard deviation 0.37 feet. A random sample of 40 pro basketball players showed they had a mean height of 6.45 feet with standard deviation 0.31. Find a 90% confidence interval for the difference of population heights.

We select option 9:2-SampleZInt since we have large samples and want to use the normal distribution. Since we have summary statistics, select Inpt: STATS. Then enter the requested values. We will use the sample standard deviations as estimates for the respective population standard deviations. The data entry requires two screens.

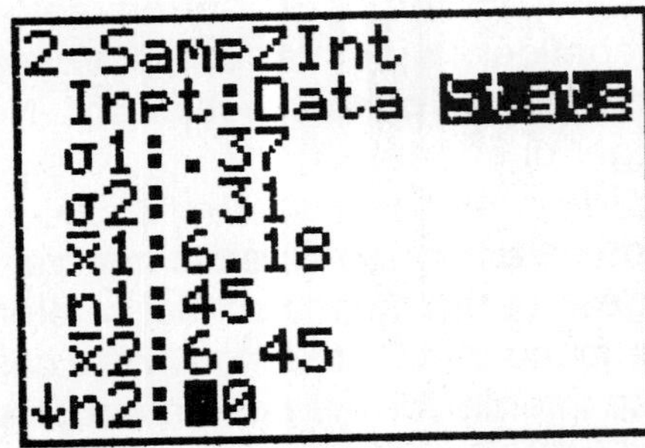


```
2-SampZInt
↑σ2:.31
x̄1:6.18
n1:45
x̄2:6.45
n2:40
C-Level:90
Calculate
```

Highlight Calculate and press ENTER.

```
2-SampZInt
 (-.3914,-.1486)
 x̄1=6.18
 x̄2=6.45
 n1=45
 n2=40
```

The interval is from -0.39 to -0.14. Since all the values are negative, we conclude that at the 90% level, basketball players mean height is greater than that of football players.

Lab Activities for Confidence Intervals for MU1- MU2 or for P1-P2

These activities coordinate with Section 8.5 of *Understandable Statistics.*

1. The following data is based on random samples of red foxes in two regions of Germany. The number of cases of rabies was counted for a random sample of 16 areas in each of two regions.

   Region 1: 10 2 2 5 3 4 3 3 4 0 2 6 4 8 7 4

   Region 2: 1 1 2 1 3 9 2 2 3 4 3 2 2 0 0 2

   a) Use a confidence level C% = 90. Does the interval indicate that the population mean number of rabies in region 1 is greater than the population mean number of rabies in region 2 at the 90% level? Why or why not?
   b) Use a confidence level C% = 95. Does the interval indicate that the population mean number of rabies in region 1 is greater than the population mean number of rabies in region 2 at the 95% level? Why or why not?
   c) Predict whether or not the population mean number of rabies in region 1 is greater than the population mean number of rabies in region 2 at the 99% level. Explain your answer. Verify your answer by running the program with a 99% confidence level

2. A random sample of 30 police officers working the night shift showed that 23 used at least 5 sick leave days per year. Another random sample of 45 police officers working the day shift showed that 26 used at least 5 sick leave days per year. Use the program CONFIDENCE INTERVAL FOR P1-P2 to find a 90%confidence interval for the difference of population proportions of police officers working the two shifts using at least 5 sick leave days per year. At the 90% level, does it seems that the proportion of officers working the night shift who use at least 5 sick leave days per year is higher than the proportion of day shift officers using that much leave? Does there seem to be a difference in proportions at the 99% level? Why or why not?

CHAPTER 9 HYPOTHESIS TESTING

The TI-83 fully supports hypothesis testing. Use the STAT key, highlight TESTS. The options used in Chapter 9 are given on the two screens.

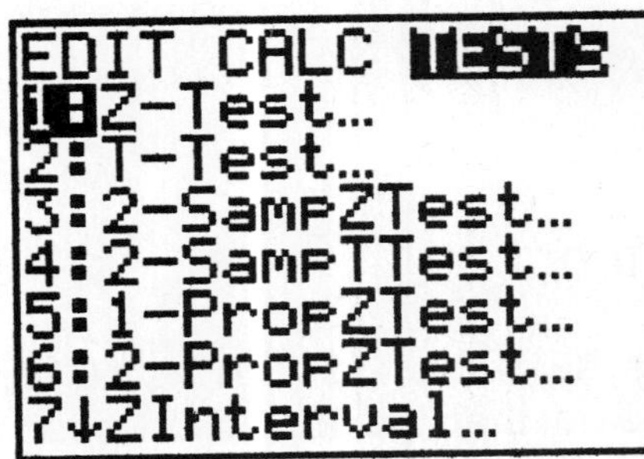

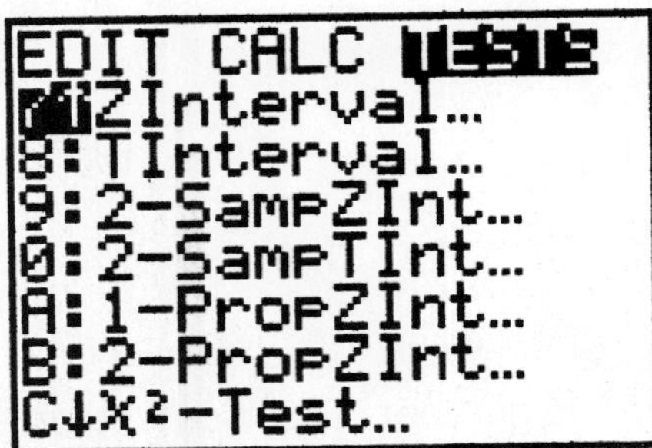

TESTING A SINGLE POPULATION MEAN
(Sections 9.1, 9.2, 9.3, 9.4 of *Understandable Statistics*)

For large sample test of a mean, we use the normal distribution. Select option **1:Z-Test** . For large samples we may estimate the population standard deviation σ by the sample standard deviation s. As with confidence intervals, we have a choice of entering raw data into a list and using the Data input option or we may use summary statistics with the Stats option. The null hypothesis is H_0: $\mu = \mu_0$. Enter the value of μ_0 in the spot indicated. To select the alternate hypothesis, use one of the three options μ: $\neq\mu_0$ $<\mu_0$ $>\mu_0$. Output consists of the value of the sample mean $\bar{x}$ and it's corresponding z value. The P-value of the sample statistic is also given.

Example (Testing a mean, large sample): Ten years ago, State College did a study regarding the number of hours full time students worked each week. At that time, the mean number of hours full time students worked per week was 8.7 hours. A recent study involving a random sample of 45 full time students showed the average number of hours worked per week was 10.3 hours with standard deviation 2.8 hours. Use a 5% level of confidence to test if the mean number of hours worked per week by full time students has increased.

Because we have a large sample, we will use the normal distribution for our sample test statistic. Select option **1:Z-Test**. We have summary statistics rather than raw data, so use the Stats input option. Enter the appropriate values on the screen.

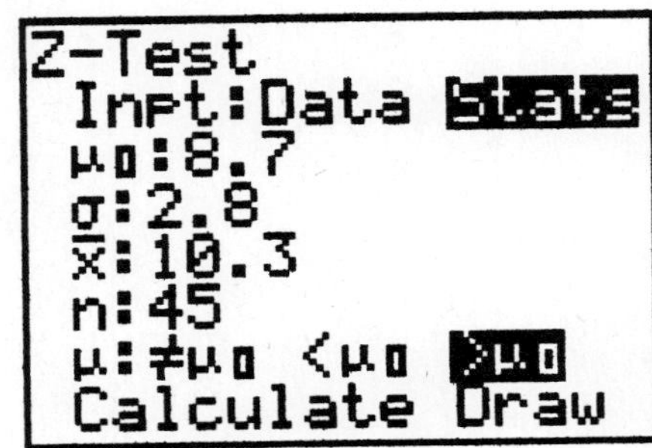

Next highlight Calculate and press Enter.

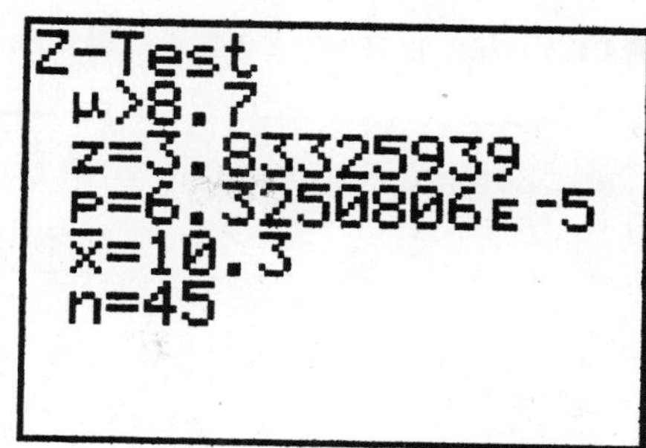

We see that the z value corresponding to the sample test statistic $\bar{x} = 10.3$ is $z = 3.83$. The critical value for a 5% level of significance and a right-tailed test is $z_0 = 1.645$. Clearly the sample z value is to the right of z_0, and we reject H_0. Notice that the P-value = 6.325E-5. This means that we move the decimal 5 places to the left, giving a P value of 0.000063. Since the P value is less than 0.05, we reject H_0.

Graph

The TI-83 gives as an option to show the sample test statistic on the normal distribution. Highlight the DRAW option on the Z-Test screen. Because the sample z is so far to the right, it does not appear on this window. However, it's value does, and a rounded P value shows as well.

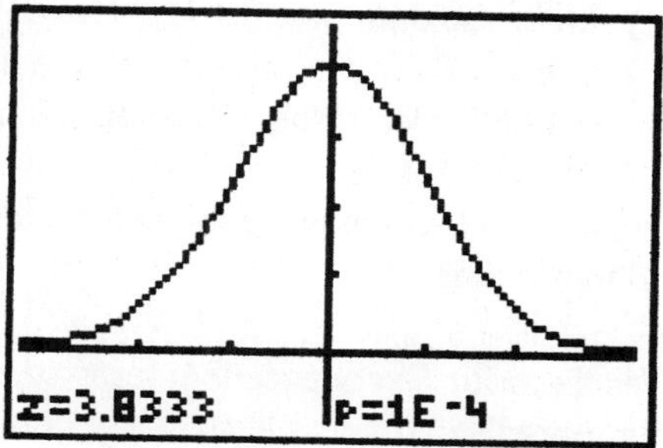

To do hypothesis testing of the mean for small samples, we use the student's t distribution. Select option **2:T-Test**. The entry screens are similar to those for the Z-test.

Lab Activities Using TESTING A SINGLE POPULATION MEAN

These activities coordinate with Sections 9.1, 9.2, 9.3 and 9.4 of *Understandable Statistics*

1. A random sample of 65 pro basketball players showed their heights (in feet) to be

6.50	6.25	6.33	6.50	6.42	6.67	6.83	6.82
6.17	7.00	5.67	6.50	6.75	6.54	6.42	6.58
6.00	6.75	7.00	6.58	6.29	7.00	6.92	6.42
5.92	6.08	7.00	6.17	6.92	7.00	5.92	6.42
6.00	6.25	6.75	6.17	6.75	6.58	6.58	6.46
5.92	6.58	6.13	6.50	6.58	6.63	6.75	6.25
6.67	6.17	6.17	6.25	6.00	6.75	6.17	6.83
6.00	6.42	6.92	6.50	6.33	6.92	6.67	6.33
6.08.							

   (a) Enter the data into list L_1. Use 1-var stat to determine the sample standard deviation.
   (b) Use the Z-Test option to test the hypothesis that the average height of the players is greater than 6.2 feet at the 1% level of significance.

2. In this problem we will see how the test conclusion is possibly affected by a change in the level of significance.

   Teachers for Excellence is interested in the attention span of students in grades 1 and 2 now as compared to 20 years ago. They believe it has decreased. Studies done 20 years ago indicate that the attention span of children in grades 1 and 2 was 15 minutes. A study sponsored by Teachers for Excellence involved a random sample of 20 students in grades 1 and 2. The average attention span of these students was (in minutes) $\bar{x} = 14.2$ with standard deviation $s = 1.5$.

   a) Use the program to conduct the hypothesis test using $\alpha = 0.05$ and a left tailed test. What is the test conclusion? What is the P value?
   b) Use the program to conduct the hypothesis test using $\alpha = 0.01$ and a left tailed test.. What is the test conclusion? How could you have predicted this result by looking at the P value from part (a)? Is the P value for this part the same as it was for part (a)?

3. In this problem, let's explore the effect that sample size has on the process of testing a mean. Run the program with the hypotheses H_0: $\mu = 200$, H_1: $\mu > 200$, $\alpha = 0.05$, $\bar{x} = 210$ and $s = 40$.

   a) Use the sample size N = 30. Note the P-value, and Z score of the sample test statistic and test conclusion.
   b) Use the sample size N = 50. Note the P-value, and Z score of the sample test statistic and test conclusion.

c) Use the sample size N = 100. Note the P-value, and Z score of the sample test statistic and test conclusion.

d) In general, if your sample statistic is close to the proposed population mean specified in H_0, and you want to reject H_0, would you use a smaller or a larger sample size?

Tests Involving a Single Proportion (Section 9.5 of *Understandable Statistics)*

To conduct a hypothesis test of a single proportion, select option **5:1-PropZTest**. The null hypothesis is H_0: $p = p_0$. Enter the value of p_0. The number of successes is designated by the value x. Enter that value. The sample size or number of trials is n. The alternate hypothesis will be prop $\neq p_0$ $< p_0$ $> p_0$. Highlight the appropriate choice. Finally highlight Calculate and press enter. Notice that the Draw option is available to show the results on the standard normal distribution.

Tests Involving Paired Differences (Dependent Samples) (Section 9.6 of *Understandable Statistics*)

The test for difference of means, dependent samples is presented in Section 9.6 of *Understandable Statistics*. Dependent samples arise from before and after studies, some studies of data taken from the same subjects, and some studies on identical twins.

To perform a paired difference test, we put our paired data into two columns, and then take the difference between the data and put the difference in another column. For example, put the before data in list L_1, the after data in list L_2. Create $L_3 = L_1 - L_2$.

Example Promoters of a state lottery decided to advertise the lottery heavily on television for one week during the middle of one of the lottery games. To see if the advertising improved ticket sales, they surveyed a random sample of 8 ticket outlets and recorded weekly sales for one week before the television campaign and for one week after the campaign. The results follow (in ticket sales) where B standard for before and A for after the advertising campaign.

B:	3201	4529	1425	1272	1784	1733	2563	3129
A:	3762	4851	1202	1131	2172	1802	2492	3151

Test the claim that the television campaign increased lottery ticket sales at the 0.05 level of significance.

We want to test to see if D = B - A is less than zero since we are testing the claim that the lottery ticket sales are greater after the television campaign.

We will put the before data in L_1, the after data inL_2, Put the differences in list L_3 by arrowing to the header L_3 and noting that it is highlighted. Type $L_3 = L_1 - L_2$ and press enter.

L1	L2	L3 3
3201	3762	-561
4529	4851	-322
1425	1202	223
1272	1131	141
1784	2172	-388
1733	1802	-69
2563	2492	71

L3 =L1−L2

Next we will conduct a t test on the differences in list L_1. Select option **2:T-Test** from the TESTS menu. Use **Data** for Inpt. Since the null hypothesis is that $\bar{d} = 0$, we use the null hypothesis H_0: $\bar{d} = \mu_0$ with $\mu_0 = 0$. Since the data is in L_3, enter that as the List. Since the test is a left-tailed test, select μ: $<\mu_0$.

Highlight Calculate and press ENTER

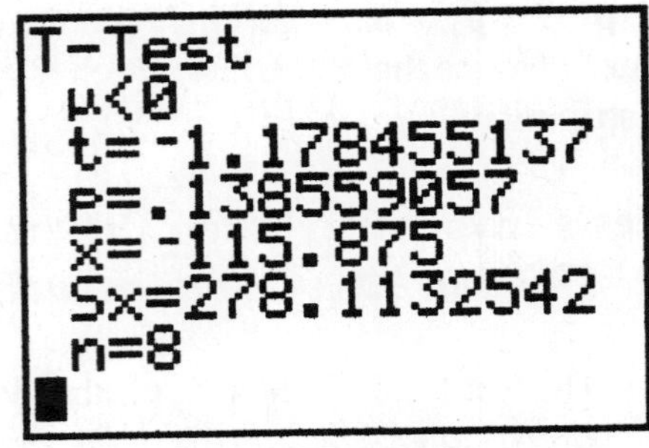

We see that the sample t value is -1.17 with a corresponding P value Of 0.138. Since the P value is greater than 0.05, we do not reject H_0.

Lab Activities Using Tests Involving Paired Differences (Dependent Samples)

1 The data are pairs of values where the first entry represents average salary ($1000/yr) for male faculty members at an institution and the second entry represents the average salary for female faculty members ($1000/yr) at the same institution. A random sample of 22 U.S. colleges and universities was used (source: Academe, Bulletin of the American Association of University Professors).

(34.5, 33.9) (30.5, 31.2) (35.1, 35.0) (35.7, 34.2) (31.5, 32.4)
(34.4, 34.1) (32.1, 32.7) (30.7, 29.9) (33.7, 31.2) (35.3, 35.5)
(30.7, 30.2) (34.2, 34.8) (39.6, 38.7) (30.5, 30.0) (33.8, 33.8)
(31.7, 32.4) (32.8, 31.7) (38.5, 38.9) (40.5, 41.2) (25.3, 25.5)
(28.6, 28.0) (35.8, 35.1)

a) Put the first entries in L_1, the second in L_2, and create L_3 to be the difference L_1 - l_2.
b) Use the **T-Test** option to test the hypothesis that there is a difference in salary. What is the P value of the sample test statistic? Do we reject or fail to reject the null hypothesis at the 5% level of significance? What about at the 1% level of significance?
c) Use the **TTest** option to test the hypothesis that female faculty members have a lower average salary than male faculty members. What is the test conclusion at the 5% level of significance? At the 1% level of significance?

2. An audiologist is conducting a study on noise and stress. Twelve subjects selected at random were given a stress test in a room that was quiet. Then the same subjects were given another stress test, this time in a room with high pitch background noise. The results of the stress tests were scores 1 through 20 with 20 indicating the greatest stress. The results follow where B represents the score of the test administered in the quiet room and A represents the scores of the test administered in the room with the high pitch background noise.

Subject	1	2	4	5	6	7	8	9	10	11	12
B	13	12	16	19	7	13	9	15	17	6	14
A	18	15	14	18	10	12	11	14	17	8	16

Test the hypothesis that the stress level was greater during exposure to high pitch background noise. Look at the P-value. Should you reject the null hypothesis at the 1% level of significance? At the 5% level?

Tests of Difference of Means (Independent Samples)
(Section 9.7 of *Understandable Statistics)*

Tests of difference of means, independent samples are presented in Sections 9.7 *Understandable Statistics*. We consider the $\overline{x}_1 - \overline{x}_2$ distribution. The null hypothesis is that there is no difference between means so H_0:$\mu_1 = \mu_2$., or H_0: $\mu_1 - \mu_2 = 0$.

Example

Sellers of microwave French fry cookers claim that their process saves cooking time. McDougle Fast Food Chain is considering the purchase of these new cookers, but wants to test the claim. Six batches of French fries were cooked in the traditional way. These times (in minutes) are

15 17 14 15 16 13

Six batches of French fries of the same weight were cooked using the new microwave cooker. These cooking times (in minutes) are

11 14 12 10 11 15

Test the claim that the microwave process takes less time. Use $\alpha = 0.05$.

Since we have small samples, we want to use the student's t distribution. Select option **4:2-SampTTest**. Put the data for traditional cooking in list L_1 and the data for the new method in list L_2.. Use Data for Inpt, use the alternate hyothesis $\mu_1 > \mu_2$, select Yes for Pooled,

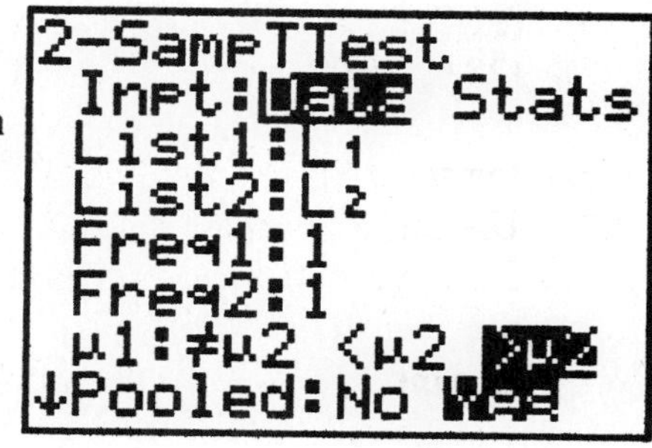

Arrow down to Calculate and highlight it. Press ENTER. The output is on two screens.

```
2-SampTTest
 μ1>μ2
 t=2.890086704
 p=.0080523896
 df=10
 x̄1=15
↓x̄2=12.16666667
```

```
2-SampTTest
 μ1>μ2
↑Sx1=1.41421356
 Sx2=1.94079022
 Sxp=1.69803808
 n1=6
 n2=6
```

Since the P value is 0.008 and it is less than 0.05, we reject H_0 and conclude that the new method cooks food faster.

Testing a Difference of Proportions
(Section 9.7 of *Understandable Statistics*)

To conduct a hypothesis test for a difference of proportions, select option **6:2-PropZTest** and enter appropriate values.

Lab Activities Using Difference of Means (Independent Samples) or Proportions

1. Calm Cough Medicine is testing a new ingredient to see if its addition will lengthen the effective cough relief time of a single does. A random sample of 15 doses of the standard medicine were tested and the effective relief times were (in minutes):

   42 35 40 32 30 26 51 39 33 28
   37 22 36 33 41

   A random sample of 20 doses were tested when the new ingredient was added. The effective relief times were (in minutes):

   43 51 35 49 32 29 42 38 45 74
   31 31 46 36 33 45 30 32 41 25

   Assume that the standard deviations of the relief times are equal for the two populations. Test the claim that the effective relief time is longer when the new ingredient is added. Use $\alpha = 0.01$.

2. Publisher's Survey did a study to see if the proportion of men who read mysteries is different than the proportion of women who read them. A random sample of 402 women showed that 112 read mysteries regularly (at least 6 books per year). A random sample of 365 men showed that 92 read mysteries regularly. Is the proportion of mystery readers different between men and women? Use a 1% level of significance.

   a) Look at the P value of the test conclusion. Jot it down.
   b) Test the hypothesis that the proportion of women who read mysteries is *greater* than the proportion of men. Use a 1% level of significance. Is the P value for a right tailed test half that of a two tailed test? If you know the P-value for a two tailed test, can you draw conclusions for a one tailed test?

CHAPTER 10 REGRESSION AND CORRELATION

LINEAR REGRESSION

(Sections 10.1, 10.2, 10.3 of *Understandable Statistics*)

Important Note: Before beginning this chapter, press CATALOG [2nd 0] and arrow down to the entry DiagnosticOn. Press ENTER twice. After doing this, the regression correlation coefficient r will appear as output with the linear regression line.

The TI-83 graphics calculator has automatic regression functions built in as well as summary statistics for two variables.

To locate the regression menu, press STAT, and then select **[CALC]**. The choice **8:LinReg(bx+a)** performs linear regression on the variables in L_1 and L_2. If the data are in other lists, then specify those lists after the LinReg(a + bx) command - separate the lists by a comma.

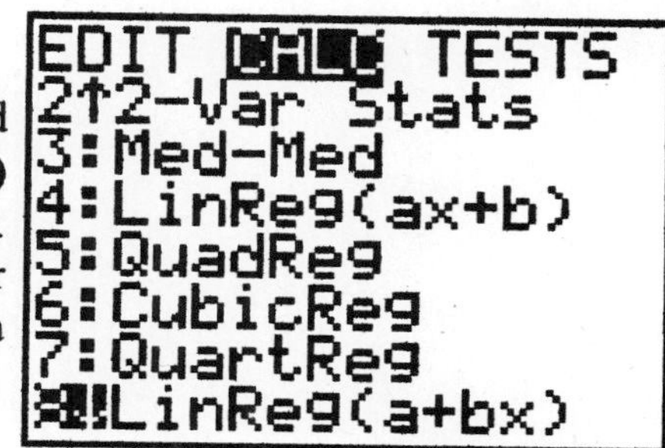

Let's look at a specific example to see how to do linear regression on the TI-83.

Example

Merchandise loss due to shoplifting, damage, and other causes is called shrinkage. Shrinkage is a major concern to retailers. The managers of H.R. Merchandise think that there is a relationship between shrinkage and number of clerks on duty. To explore this relationship, a random sample of 7 weeks was selected. During each week the staffing level of sales clerks was kept constant and the dollar value of the shrinkage was recorded.

# Sales Clerks:	12	11	15	9	13	8
Shrinkage:	15	20	9	25	12	31

a) Find the equation of the least squares line with number of sales clerks as the explanatory variable.

First put the number of sales clerk data in L_1 and the corresponding amount of shrinkage in L_2.

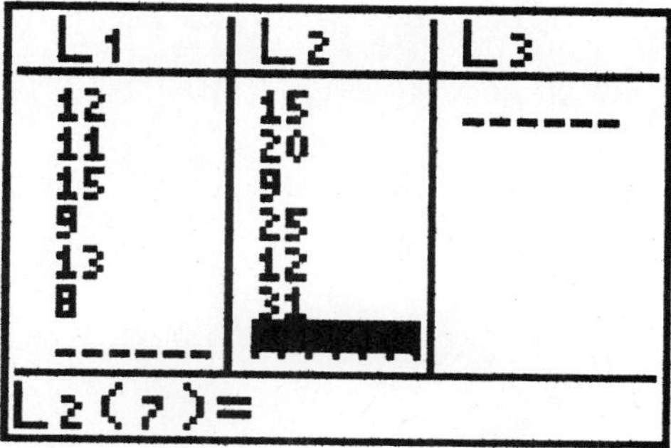

Use the **CALC** menu and select **8:LinReg(a+bx).** Press ENTER.

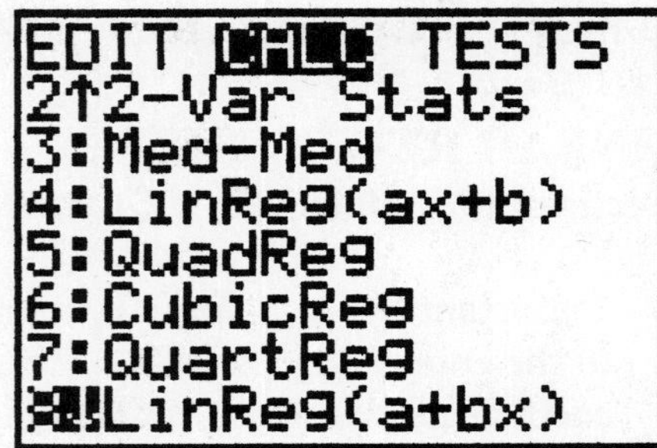

Designate the lists containing the data. Then press ENTER

```
LinReg(a+bx) L1,
L2
```

Note that the linear regression equation is

y = 54.48x - 3.16

Also we have the value of the Pearson correlation coefficient given.

r = -0.982 and the value of r^2.

```
LinReg
 y=a+bx
 a=54.48
 b=-3.16
 r²=.9638610039
 r=-.9817642303
```

b) Make a scatter plot and show the least squares line on the plot.

To graph the least squares line, press .

Clear all existing equations from the display. To enter the least squares line equation in directly, press VARS. Select item **5:Statistics**.

```
VARS Y-VARS
1:Window...
2:Zoom...
3:GDB...
4:Picture...
5:Statistics...
6:Table...
7:String...
```

Press **ENTER** and then select **EQ** and then **1:RegEQ.**

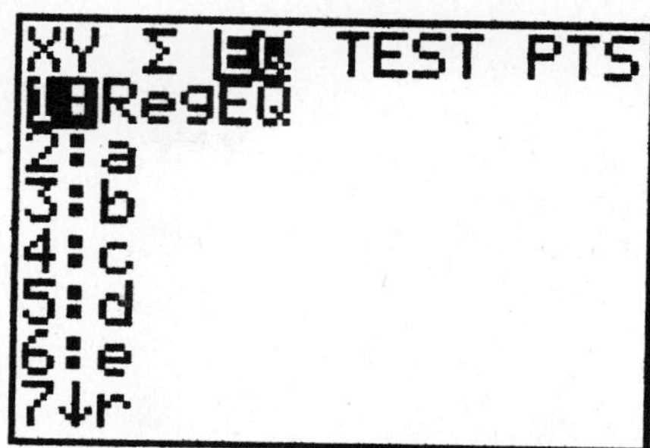

You will see the least squares regression equation appear automatically into the graphing equations list.

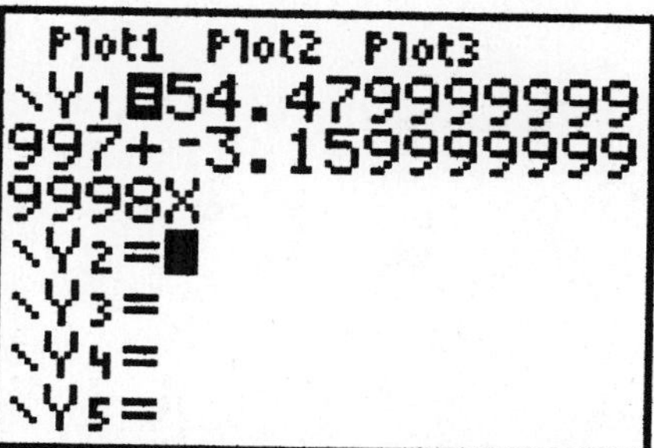

We want to graph the least squares line with the scatter plot. Next press 2ND [**STAT PLOT**] and select **1:Plot1**. Highlight **ON**, **Scatter plot picture,** $\mathbf{L_1}$ for Xlist and $\mathbf{L_2}$ for Ylist.

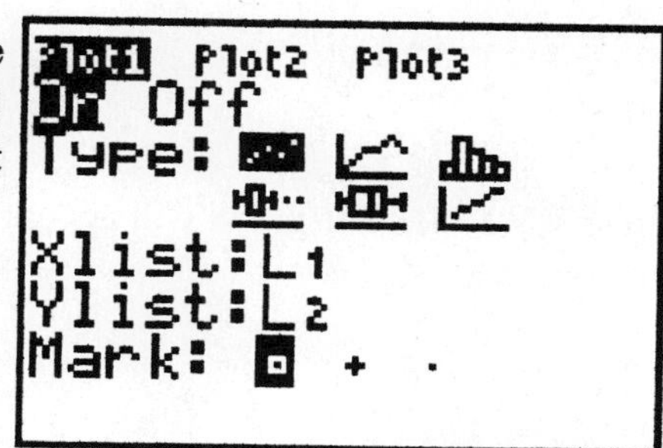

The last step before we graph is to set the graphing window. Press WINDOW. Set Xmin to be lowest x value, Xmax the highest x value, ymin at -5 and ymax at above highest value.

```
WINDOW
 Xmin=8
 Xmax=15
 Xscl=1
 Ymin=-5
 Ymax=30
 Yscl=1
 Xres=1
```

Now press GRAPH.

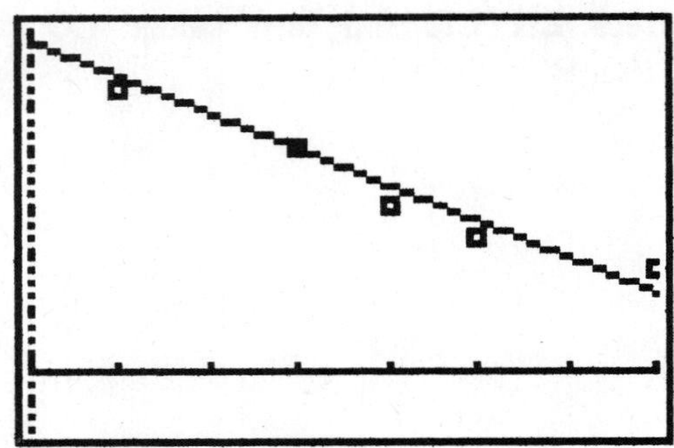

c) Predict the shrinkage when 10 clerks are on staff.

Press 2ND **[CALC].** This is the Calculate function for graphs and is different from the CALC you highlight from the STAT menu. Highlight **1:Value.**

Press **ENTER.** The graph will appear. Enter 10 next to **EVAL X=**.

Press **ENTER.** The predicted value for Y appears.

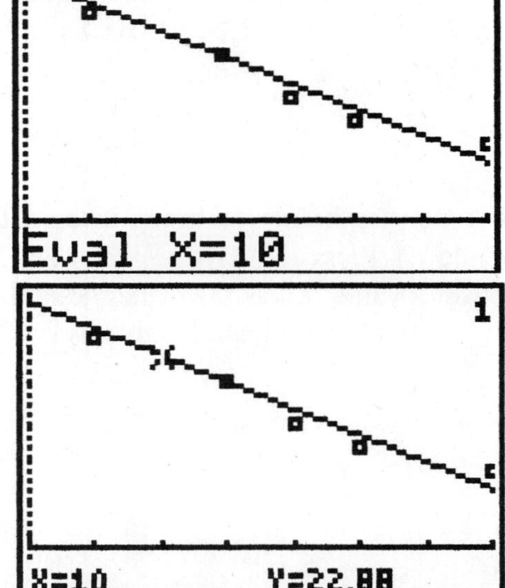

To display the summary statistics for X and Y, press STAT, highlight **CALC.** Select **2:2-Var Stats.** Press ENTER

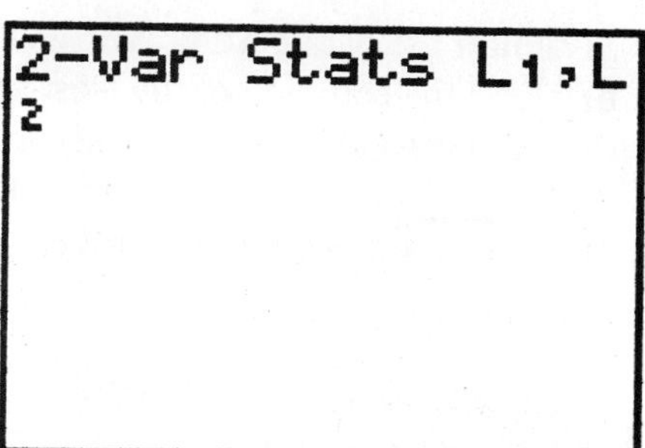

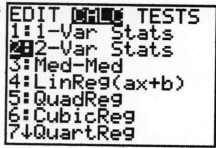

Type the lists containing the data and press ENTER

```
2-Var Stats L1,L
2
```

Arrow to the statistics of interest.

```
2-Var Stats
 x̄=11.33333333
 Σx=68
 Σx²=804
 Sx=2.581988897
 σx=2.357022604
↓n=6
```

```
2-Var Stats
↑Σy=112
 Σy²=2436
 Sy=8.310635756
 σy=7.586537784
 Σxy=1164
↓minX=8
```

Lab Activities for Linear Regression and Correlation

These lab activities coordinate with Sections 10.1, 10.2, and 10.3 of *Understandable Statistics*.

For each of the following data sets do the following:

a) Enter the data putting X values in L_1 and Y values in L_2.
b) Find the equation of the least squares line.
c) Find the value of the correlation coefficient r.
d) Draw a scatter plot and show the least squares line on the scatter plot.
e) Use 2-Var Stats to find all the sums necessary for computing the standard error of estimate S_e.

1. **CRICKET CHIRPS VERSUS TEMPERATURE**

IN THE FOLLOWING DATA PAIRS (X,Y)

X = CHIRPS/SEC FOR THE STRIPED GROUND CRICKET

Y = TEMPERATUTE IN DEGREES FAHRENHEIT

SOURCE: THE SONG OF INSECTS BY DR. G.W. PIERCE
HARVARD COLLEGE PRESS

(20.000, 88.600)	(16.000, 71.600)	(19.800, 93.300)
(18.400, 84.300)	(17.100, 80.600)	(15.500, 75.200)
(14.700, 69.700)	(17.100, 82.000)	(15.400, 69.400)
(16.200, 83.300)	(15.000, 79.600)	(17.200, 82.600)
(16.000, 80.600)	(17.000, 83.500)	(14.400, 76.300)

This is the data in Table 10-3, Section 10.2 of *Understandable Statistics*.

2. **LIST PRICE VERSUS BEST PRICE FOR A NEW GMC PICKUP TRUCK**

IN THE FOLLOWING DATA PAIRS (X,Y)

X = LIST PRICE (IN $1000) FOR A GMC PICKUP TRUCK
Y = BEST PRICE (IN $1000) FOR A GMC PICKUP TRUCK

SOURCE: CONSUMERS DIGEST, FEBRUARY 1994

Lab Activities for Linear Regression and Correlation continued

(12.400, 11.200)	(14.300, 12.500)	(14.500, 12.700)
(14.900, 13.100)	(16.100, 14.100)	(16.900, 14.800)
(16.500, 14.400)	(15.400, 13.400)	(17.000, 14.900)
(17.900, 15.600)	(18.800, 16.400)	(20.300, 17.700)
(22.400, 19.600)	(19.400, 16.900)	(15.500, 14.000)
(16.700, 14.600)	(17.300, 15.100)	(18.400, 16.100)
(19.200, 16.800)	(17.400, 15.200)	(19.500, 17.000)
(19.700, 17.200)	(21.200, 18.600)	

3. **DIAMETER OF SAND GRANULES VERSUS SLOPE ON A NATURAL OCCURRING OCEAN BEACH**

IN THE FOLLOWING DATA PAIRS (X,Y)

X = MEDIAN DIAMETER (MM) OF GRANULES OF SAND
Y = GRADIENT OF BEACH SLOPE IN DEGREES

THE DATA IS FOR NATURALLY OCCURRING OCEAN BEACHES.

SOURCE: PHYSICAL GEOGRAPHY BY A.M. KING
OXFORD PRESS , ENGLAND

(0.170, 0.630)	(0.190, 0.700)	(0.220, 0.820)
(0.235, 0.880)	(0.235, 1.150)	(0.300, 1.500)
(0.350, 4.400)	(0.420, 7.300)	(0.850, 11.300)

See Chapter 10 Using Technology of *Understandable Statistics* for a more complete discussion of this data.

4. **NATIONAL UNEMPLOYMENT RATE MALE VERSUS FEMALE**

IN THE FOLLOWING DATA PAIRS (X,Y)

X = NATIONAL UNEMPLOYMENT RATE FOR ADULT MALES
Y = NATIONAL UNEMPLOYMENT RATE FOR ADULT FEMALES

SOURCE: STATISTICAL ABSTRACT OF THE UNITED STATES

(2.900, 4.000)	(6.700, 7.400)	(4.900, 5.000)
(7.900, 7.200)	(9.800, 7.900)	(6.900, 6.100)
(6.100, 6.000)	(6.200, 5.800)	(6.000, 5.200)
(5.100, 4.200)	(4.700, 4.000)	(4.400, 4.400)
(5.800, 5.200)		

CHAPTER 11 CHI SQUARE AND F DISTRIBUTIONS

CHI SQUARE TEST FOR INDEPENDENCE
(Section 11.1 of *Understandable Statistics*)

The TI-83 calculator supports tests for independence. Press the STAT key, select TESTS, and the use option **C: χ^2-Test...** The original observed values need to be entered in a matrix.

Example A computer programming aptitude test has been developed for high school seniors. The test designers claim that scores on the test are independent of the type of school the student attends: rural, suburban, urban. A study involving a random sample of students from each of these types of institutions yielded the following information where aptitude scores range from 200-500 with 500 indicating the greatest aptitude and 200 the least. The entry in each cell is the observed number of students making the indicated score on the test.

School Type

Score	Rural	Suburban	Urban
200 - 299	1 33	2 65	3 82
300 - 399	4 45	5 79	6 95
400 - 500	7 21	8 47	9 63

Using the option **C: χ^2-Test...**, test the claim that the aptitude test scores are independent of the type of school attended at the 0.05 level of significance.

We need to use a matrix to enter the data. Press the MATRIX key. Highlight Edit, and select1. [A]. Press ENTER

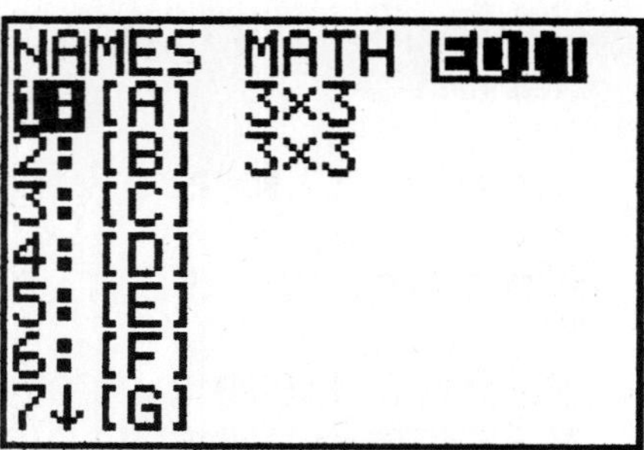

Since the contingency table has 3 rows and 3 columns, type 3 (for number of rows), press ENTER, type 3 (for number of columns), press ENTER. Then type in the observed values. Press ENTER after each entry. Notice that you enter the table by rows. When the table is entered completely, press the MATRIX key again.

This time, select 2: [B] Set the dimensions to be the same as for matrix [A]. For this example, [B] should be 3 rows×3 columns.

Now press STAT key, highlight TESTS, and use option **C: χ^2-Test...** Press ENTER This tells you that the observed values of the table are in matrix [A]. The expected values are in corresponding positions of matrix [B]. Note: If you want to see matrix [B], press the MATRIX key, under NAMES select 2:[B]. Press ENTER twice.

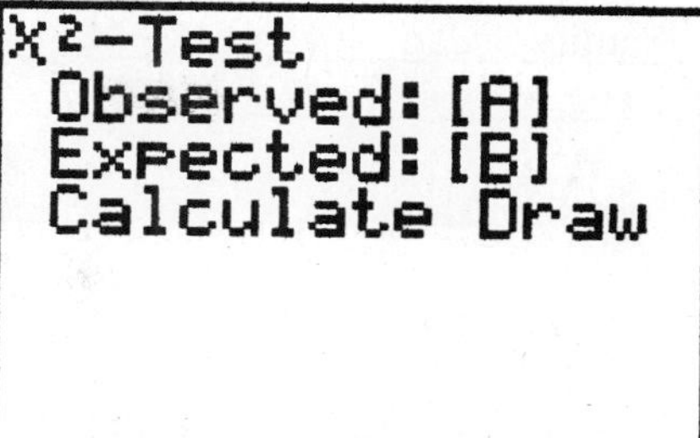

Highlight Calculate and press ENTER . We see that the sample value of χ^2 is 1.32. The P value is 0.857. Since the P value is larger than 0.05, we do not reject the null hypothesis.

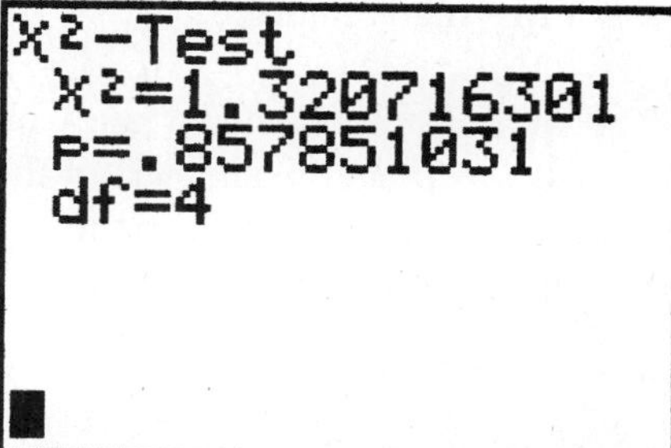

Graph

Notice that one of the options of the χ^2-Test is to graph the χ^2 disribution and show the sample test statistic on the graph. Highlight Draw and press ENTER. Before you do this, be sure that all STAT PLOTs are set to OFF, and that you have cleared all the entries in the Y= menu.

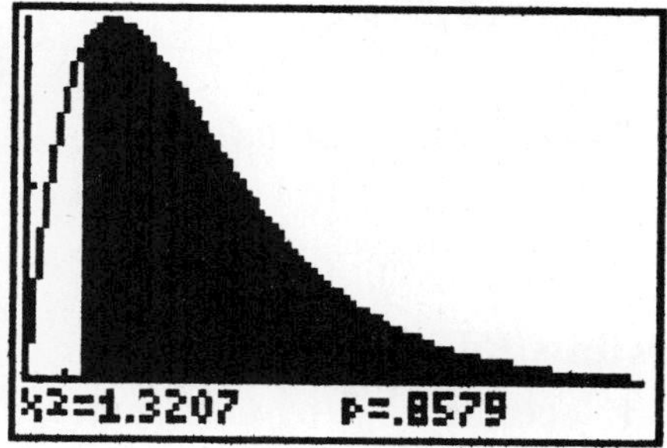

Lab Activities for CHI SQUARE TEST OF INDEPENDENCE

These activities coordinate with Section 11.1 of *Understandable Statistics*.

1. We Care Auto Insurance had its staff of actuaries conduct a study to see if vehicle type and loss claim are independent. A random sample of auto claims over the 1st six months give the information in the contingency table.

Total Loss Claims per Year per Vehicle

Type of vehicle	$0-999	$1000-2999	$3000-5900	$6000+
Sports car	20	10	16	8
Truck	16	25	33	9
Family Sedan	40	68	17	7
Compact	52	73	48	12

Test the claim that car type and loss claim are independent. Use $\alpha = 0.05$.

2. An educational specialist is interested in comparing three methods of instruction:

   S.L.- standard lecture with discussion
   T.V.- video taped lectures with no discussion
   I.M.- individualized method with reading assignments and tutoring, but no lectures.

   The specialist conducted a study of these three methods to see if they are independent. A course was taught using each of the three methods and a standard final exam was given at the end. Students were put into the different method sections at random. The course type and test results are shown in the contingency table.

Final Exam Score

Course Type	below 60	60-69	70-79	80-89	90-100
S.L.	10	4	70	31	25
T.V.	8	3	62	27	23
I.M.	7	2	58	25	22

Test the claim that the instruction method and final exam test scores are independent using $\alpha = 0.01$.

Testing Two Variances
(Section 11.5 of *Understandable Statistics)*

Under the TESTS menu of the STAT key, option **D: 2-SampFTest** provides hypothesis testing for two variances. Again there is a choice of data entry styles: raw data in lists or summary statistics for which you provide the sample standard deviations and sample sizes.

Example: Two manufacturing processes for making paint are under study. A random sample of 35 applications of the paint produced under the first method shows the duration has a standard deviation of 1.5 years. For the second method, the a random sample of 40 applications shows a standard deviation of 1.3 years. Use a 5% level of significance to test if the variances are equal.

Use the **D: 2-SampFTest** option. Select Inpt: Stats and enter the data as requested. Note that we follow the convention of always entering the larger standard deviation as s_1.

Highlight Calculate and press ENTER. We see the sample F statistic is 1.33 with P-value 0.386. We do not reject H_0.

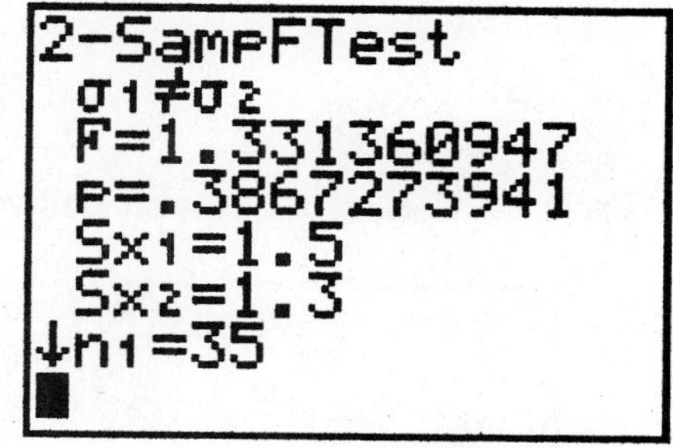

One-Way ANOVA
(Section 11.5 of *Understandable Statistics*)

The TI-83 supports one-way ANOVA tests. Press the STAT key, under TESTS select option **F:ANOVA(**. First the data are entered into lists, one list for each treatment. The format of the command for ANOVA is **ANOVA(L_1, L_2, etc)** where all the treatment lists are given.

Example A psychologist has developed a series of tests to measure depression level. The composite scores range from 50 to 100 with 100 representing the most severe depression level. This measuring device was used in a study of treatments for depression. A random sample of 12 patients with approximately the same depression level as measured by the tests was divided into 3 different treatment groups. Then, one month after treatment was competed, the depression level of each patient was again evaluated using the series of tests. The after treatment depression levels are given

Treatment 1: 70 65 82 83 71
Treatment 2: 75 62 81
Treatment 3: 77 60 80 75

Use the option **F:ANOVA(** to test the claim that the population means are all the same at the 5% level of significance.

Enter the treatment data in lists L_1, L_2, L_3 respectively.

Select the F:ANOVA(option from the TESTS menu of the STAT key. Press ENTER. Then type the names of the 3 lists containing the data. Separate the list names by commas.

Press ENTER. The output is on two screens.

```
One-way ANOVA
 F=.0362010431
 p=.9645860759
 Factor
  df=2
  SS=5.45
↓ MS=2.725
```

```
One-way ANOVA
↑ MS=2.725
 Error
  df=9
  SS=677.466667
  MS=75.2740741
 Sxp=8.67606328
```

We see the sample F statistic is F = 0.036 with P-value 0.9645. We do not reject H_0 and conclude that there is no evidence that some of the means are different.

Lab Activities for ANALYSIS OF VARIANCE

These activities coordinate with section 11.4 of *Understandable Statistics.*

1 A random sample of 20 overweight adults were randomly divided into 4 groups. Each group was given a different diet plan, and the weight loss for each individual after 3 months follows:

Plan 1: 18 10 20 25 17
Plan 2: 28 12 22 17 16
Plan 3: 16 20 24 8 17
Plan 4: 14 17 18 5 16

Test the claim that the population mean weight loss is the same for the four diet plans at the 5% level of significance.

2. A psychologist is studying the time it takes rats to respond to stimuli after being given doses of different tranquilizing drugs. A random sample of 18 rats were divided into 3 groups. Each group was given a different drug. The response time to stimuli was measured (in seconds). The results follow.

Drug A 3.1 2.5 2.2 1.5 0.7 2.4
Drug B 4.2 2.5 1.7 3.5 1.2 3.1
Drug C 3.3 2.6 1.7 3.9 2.8 3.5

Test the claim that the population mean response times for the three drugs is the same at the 5% level of significance.

3. A research group is testing various chemical combinations designed to neutralize and buffer the effects of acid rain on lakes. A random sample of 18 lakes of similar size in the same region have all been affected in the same way by acid rain. The lakes are divided into four groups and each group of lakes is sprayed with a different chemical combination. An acidity index is then taken after treatment. The index ranges from 60 to 100 with 100 indicating the greatest acid rain pollution. The results follow.

Combination I 63 55 72 81 75
Combination II 78 56 75 73 82
Combination III 59 72 77 60
Combination IV 72 81 66 71

Test the claim that the population mean acidity index after each of the four treatments is the same at the 0.01 level of significance.

PART II

COMPUTERSTAT VERSION 5

FOR

UNDERSTANDABLE STATISTICS
SIXTH EDITION

OR

UNDERSTANDING BASIC STATISTICS

CHAPTER 1 GETTING STARTED

ABOUT COMPUTERSTAT

ComputerStat Version 5 is a collection of computer programs designed for beginning statistics students and class demonstrations. The software is available in Windows (runs with Windows 3.1, 95, 98), Macintosh®, as well as IBM® (and Compatible),DOS format. The programs together with this *Guide* constitute an educational tool rather than a commercial or research tool. The programs are designed to demonstrate statistical concepts as well as to process data.

There are more than 50 Class Demonstrations available in *ComputerStat Version 5*. Most of these Class Demonstrations use real data from designated references. A list of all the data files included in Class Demonstrations can be found in the Appendix of this *Guide*.

The programs in *ComputerStat* are interactive and require minimal experience using computers, as well as minimal computer devices. With the briefest introduction to the basic processes of using a computer, students can use *ComputerStat*. The Lab Activities in this *Guide* support active classroom discussion. The *Guide* and programs are written so that very little classroom time need be devoted to learning how to use *ComputerStat*. Students also can work in a computer lab and gain useful insights from working with *ComputerStat* and the Lab Activities.

ComputerStat is not a tutorial package. Rather, it gives students an opportunity to further explore concepts they mastered in class without the necessity of doing a lot of calculations by hand.

Some advantages of *ComputerStat* are:

- *ComputerStat* is easy to use and requires minimal equipment. No data files need to be created. Class Demonstrations that can be selected from a menu contain real data from referenced sources. Students also may select to enter their own data directly as instructed by the program.

- The programs and Lab Activities are directly related to the corresponding units of *Understandable Statistics* fifth edition.

- The programs together with the Lab Activities help students explore what happens when various conditions of a problem are modified.

- *ComputerStat* is inexpensive, does not require extensive equipment, and does not require extensive time to learn to use.

Hardware and Software Requirements for

Windows Based Computers

IBM® compatibility
80386, 80486, or higher CPU
4 Mb of available RAM
Hard Disk drive and 1 high-density floppy drive
Microsoft Windows 3.1 or 95 or 98
VGA or SVGA video adapter
Color or monochrome monitor

Macintosh Computers

ComputerStat for Macintosh will run on most Macintosh systems, beginning with the Mac SE.

2 Mb of available RAM
Hard disk drive and 1 high-density floppy drive
System 6.0.7 or higher
Color or monochrome monitor, 9" or larger

DOS Based Computers

IBM ® compatibility
8088 or higher micro-processor chip
640K bytes of available RAM
1 high density diskette drive
MS-DOS® 3.1 or higher
CGA, EGA, MCGA, VGA, or SVGA video adapter
Color or monochrome monitor

ComputerStat is designed to operate with or without printers. Students can see results on the screen, and when they have studied them sufficiently, they can tell the computer to continue with the program.

Installing *CompuerStat* on Windows Based Computers

To set up the program in Windows, first make a copy of the *ComputerStat* diskettes. Install *ComputerStat* on your hard drive by the following procedure. Place your working copy of *ComputerStat* in your floppy drive. From the Windows Program press **Start** and select **Run**. Type **A:\setup** (where "A: is theletter of your disk drive). Press the Enter key and follow the instructions on-screen.

Starting *ComputerStat* Windows Version

To start, simply double-click the *ComputerStat* icon on your hard drive. Select the icon that lists the drive you prefer to use as the default drive.

Use the menu and follow the instructions on the screen.

Installing *ComputerStat* Version 5 on Macintosh Computers

Install *ComputerStat* on your hard drive according to the following procedure.

1. Create a folder on your hard drive to hold the *ComputerStat* files. Name this folder COMPUTERSTAT.
2. Insert the *CompuerStat* disk in the floppy dirve and open it by double-clicking on the disk icon.
3. Go tothe Edit menu and choose Select All to select thefiles in the *CoomputerStat* disk window.
4. Drag the disk files to the folder you have created on your hard drive.
5. Eject the *ComputerStat* disk from your floppy drive by dragging its icon to the trash can.
6. Repeat steps 2-5 for any additional disks (if applicable).

Starting *ComputerStat* on Macintosh Computers

To start, simply double-click the *ComputerStat* program icon on your hard drive

Use the menu and follow the instructions on the screen.

Installing *ComputerStat* Version 5 on DOS Based Computers

Make a copy of the *ComputerStat* diskette. If you have a hard drive, make a directory and copy the *ComputerStat* files from the diskette to the hard drive. Otherwise, use the copy of *ComputerStat* in a diskette drive.

Starting *ComputerStat* on DOS Based Computers

At the DOS prompt, change to the directory containing *ComputerStat*.

Type **CSTAT** at the prompt.

For instance, if the *ComputerStat* diskette is in drive A, then change to drive A and type CSTAT.

C> **A:**
A> **CSTAT**

Use the menu and follow the instructions on the screen

USING *ComputerStat* Version 5

After starting *ComputerStat* you will see opening title screens. Follow the instructions and click on the NEXT button until the main menu appears.

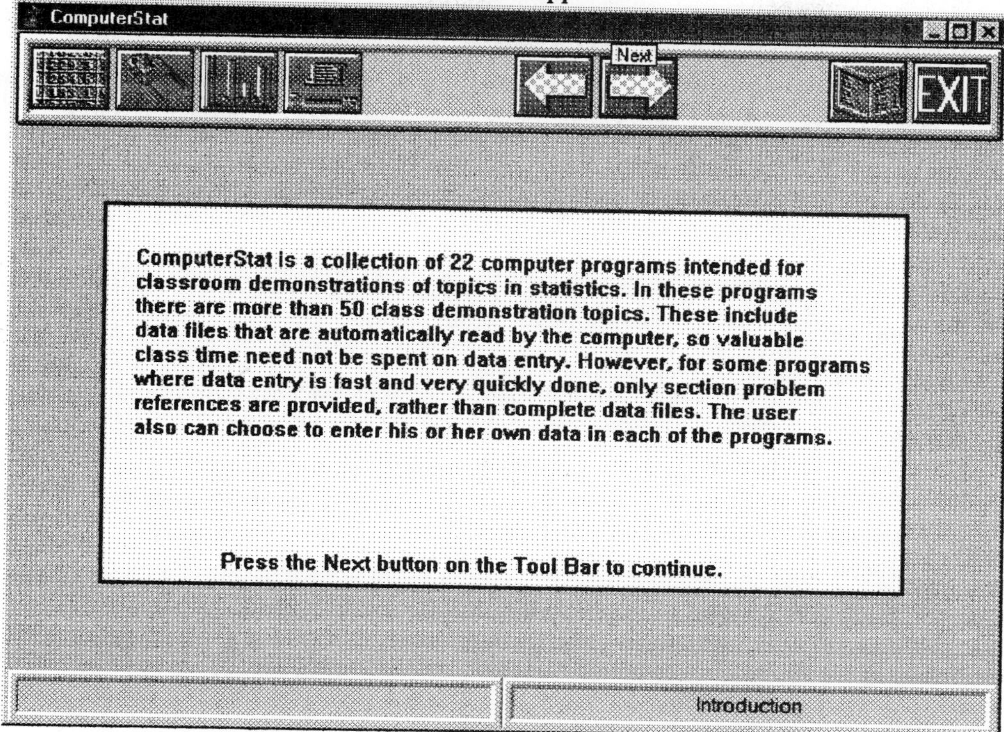

ComputerStat
Next
EXIT
ComputerStat is a collection of 22 computer programs intended for classroom demonstrations of topics in statistics. In these programs there are more than 50 class demonstration topics. These include data files that are automatically read by the computer, so valuable class time need not be spent on data entry. However, for some programs where data entry is fast and very quickly done, only section problem references are provided, rather than complete data files. The user also can choose to enter his or her own data in each of the programs.
Press the Next button on the Tool Bar to continue.
Introduction

Main Menu

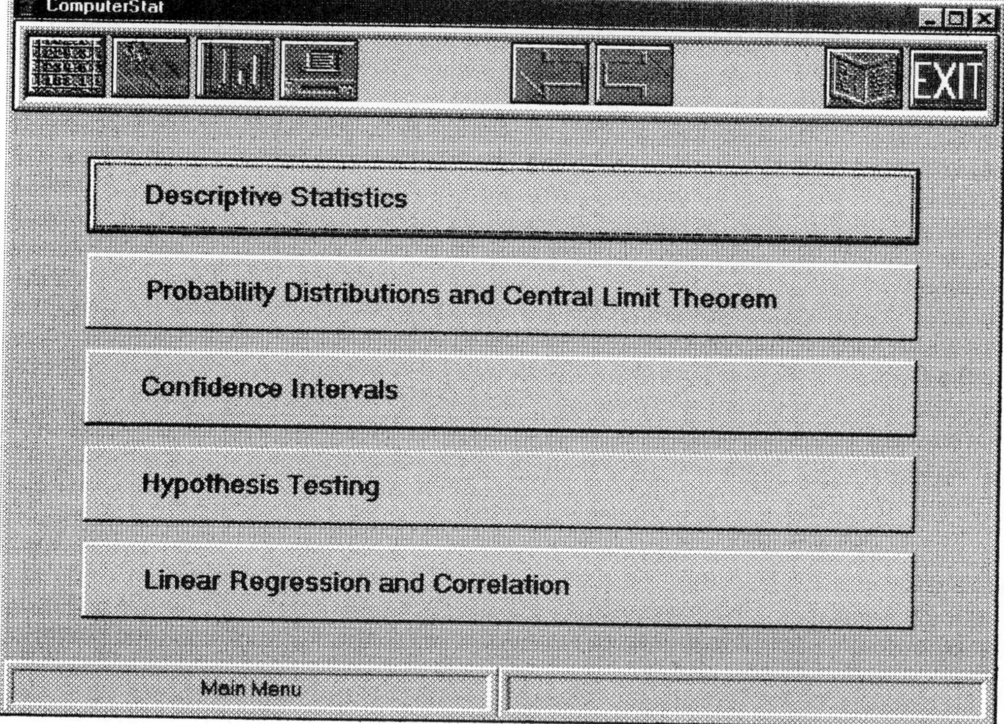

Communicating with the Computer

The computer keyboard works very much like a typewriter. In addition, there are some special keys. As you use *ComputerStat*, you will find that in most cases the computer wants you to respond with a number.

Every time you wish to type a number, use the key with the desired digit. The period serves as a decimal point. Always use the number "Zero" and not the letter key "Oh". Also, be sure to use the number 1 and not the lower case of the letter "L".

Using the Mouse: Generally you will left click the mouse to activate a menu item or to activate a response box.

Notation on the Computer

Some mathematical symbols are expressed in a slightly different format on a computer.

Mathematical Notation	Computer Notation
≤ (less than or equal to)	<=
≥ (greater than or equal to)	>=
≠ (not equal)	<>

Occasionally numbers will be expressed in E notation such as 3.012758E-02. This notation tells us we are to multiply the number in front of the E by the power of 10 listed after the E. For instance

3.012758E-02 = 0.03012758	The E-02 tells us to multiply 3.012758 by 10^{-2}. This has the effect of moving the decimal point two places to the left.
5.738613E+03 = 5738.613	The E+03 part tells us to multiply 5.728613 by 10^3. This has the effect of moving the decimal three places to the right.

In short, if the value following E is

negative, move the decimal that many places to the *left.*

positive, move the decimal that many places to the *right*.

Special note for DOS based computers: If the computer is expecting a number and you press the ← ENTER or return key without typing a number, the computer thinks you entered *zero*.

CHAPTER 2 ORGANIZING DATA

RANDOM SAMPLES

(Section 2.1 Random Samples of *Understandable Statistics*)

Main menu selection: DESCRIPTIVE STATISTICS

Sub-menu selection: RANDOM SAMPLES

I. Description of the program

In this program a random number generator is used to do the following activities

a) Select a random sample of size M from a population of size N and list the chosen sample items. You may sample *with* or *without* replacement.

b) Simulate the experiment of tossing one die up to 5000 times

c) Simulate the experiment of tossing two dice up to 5000 times

Input: First you select the option you wish to run.

Option 1: Select a random sample of size M from a population of size N.
Population size N; N between 1 and 1,000,000
Sample size M; M between 1 and 500 for sampling with replacement;
M between 1 and the minimum of N or 500 for sampling without replacement..
Choices available: sampling with or without replacement

Option 2: Simulate the experiment of tossing one die.
Number of times to toss die, N: N between 1 and 5000

Option 3: Simulate the experiment of tossing two dice.
Number of time to toss dice, N; N between 1 and 5000

Finally you have a choice of repeating any of the three options , returning to the sub-menu, or returning to the main menu.

Output: For the three options

Option 1: The number corresponding to the data elements from the population that are to be included in the sample.

Option 2: Number of occurrences of each of the 6 possible outcomes of tossing one die and the relative frequencies of the outcomes.

Option 3: Number of occurrences of each of the 36 possible outcomes of tossing two dice and the corresponding relative frequencies.
Number of occurrences of each of the 12 possible sums on the two dice with corresponding relative frequencies.

II. Sample Run

Suppose there are 75 students enrolled in a large section of statistics. Draw a random sample of 15 students without using any student in the sample more than once.

First assign each of the 75 students a number from 1 to 75. The population size is 75. The sample size is 15. Instruct the computer to sample without replacement since you want each of the students in the sample to be different. The computer will give you a list of numbers. These are the numbers assigned to the students to be included in your sample. From the student numbers, you can determine which students are in your sample.

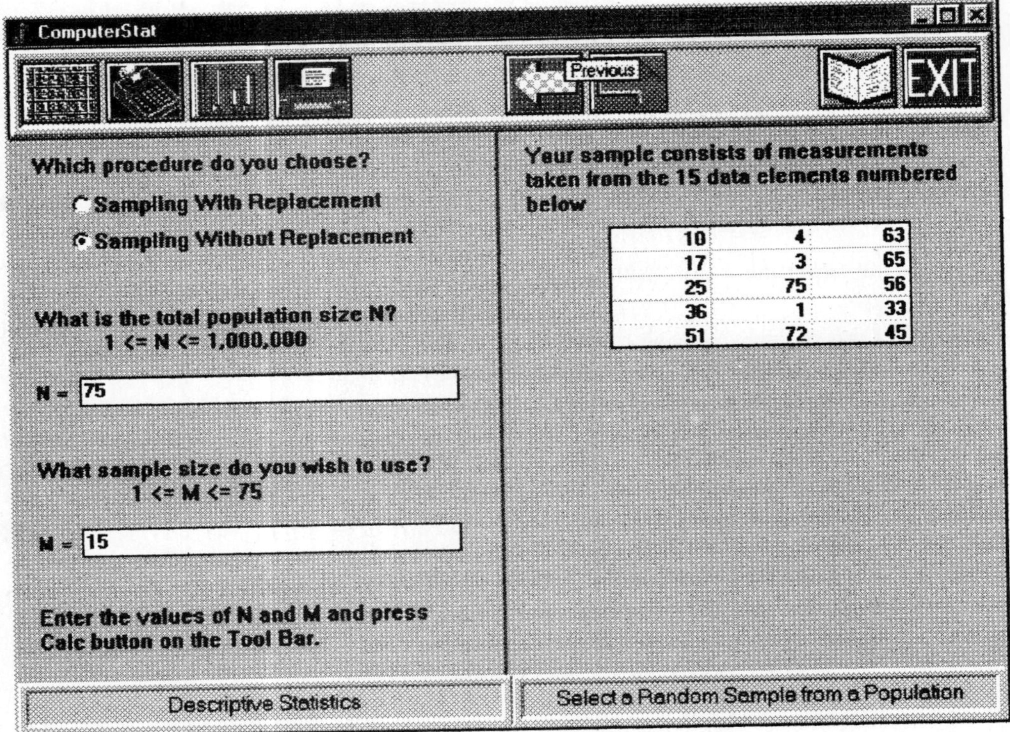

Lab Activities Using RANDOM SAMPLES

These activities coordinate to Section 2.1 Random Samples in the text *Understandable Statistics*

1. Out of a population of 8000 eligible county residents, select a random sample of 50 for prospective jury duty. Should you sample with or without replacement? Explain.

Simulating experiments in which outcomes are equally likely is another important use of random numbers.

2. We can simulate dealing bridge hands by numbering the cards in a bridge deck from 1 to 52. Then we draw a random sample of 13 numbers without replacement from the population of 52 numbers. A bridge deck has 4 suits: Hearts, diamonds, clubs, and spades. Each suit contains 13 cards; those numbed 2 through 10 a jack, a queen, a king, and an ace. Decide how to assign the numbers 1 through 52 to the cards in the deck. Use the ComputerStat program RANDOM SAMPLES to get the numbers of the 13 cards in one hand. Translate the numbers to specific cards and tell what cards are in the hand. For a second game the cards would be collected and reshuffled. Rerun the program option and determine the hand you might get in a second game.

3. We can also simulate the experiment of tossing a fair coin. The possible outcomes resulting from tossing a coin are heads or tails. Assign the outcomes heads the number 2 and the outcome tails the number 1. Use the ComputerStat program RANDOM SAMPLES to simulate the act of tossing a coin 10 times. Note that you will use a population size of 2 since that is the number of outcomes. You need to sample with replacement. Select a sample of size 10 since you will be tossing the coin 10 times. Record the numbers of the outcomes and then record the outcomes themselves. Repeat the process for 30 trials.

4. When we toss a die, there are six possible outcomes: the number of dots appearing on the top will be 1 or 2 or 3 or 4 or 5 or 6. If the die is fair, each of the outcomes is equally likely to occur. However, if you toss the die 6 times, you will not necessarily get all six outcomes. It is quite possible to get two or three of the outcomes several times, and not get other outcomes at all. Run RANDOM SAMPLES and select Option 2 which simulates the experiment of tossing one die. Record the outcomes of tossing the die N = 6 times. Are the outcomes fairly evenly distributed? When you study probability (Chapter 4 of the text *Understandable statistics*) you will gain an understanding of why the outcomes of the experiments can vary. We will return to this program then and use the dice experiments to gain insights into probability.

FREQUENCY DISTRIBUTIONS

(Section 2.3 Histograms and Frequency Distributions of *Understandable Statistics*)

Main menu selection: DESCRIPTIVE STATISTICS

Sub-men selection: FREQUENCY DISTRIBUTIONS AND GROUPED DATA

I. Description of the program

This program produces a frequency distribution, relative frequency distribution, and cumulative frequency distribution with up to 10 class. In addition it shows the corresponding histogram and the corresponding ogive. The frequency table shows class boundaries, class frequencies, relative frequencies, class midpoints, and cumulative frequencies. In addition, there is an option for computing the mean and standard deviation both from the raw data and from the grouped data.

Input: Option I: Use an existing class demonstration data file that is included in the data menu of *ComputerStat*. The data values included in the demonstrations are displayed in Appendix I of this guide.

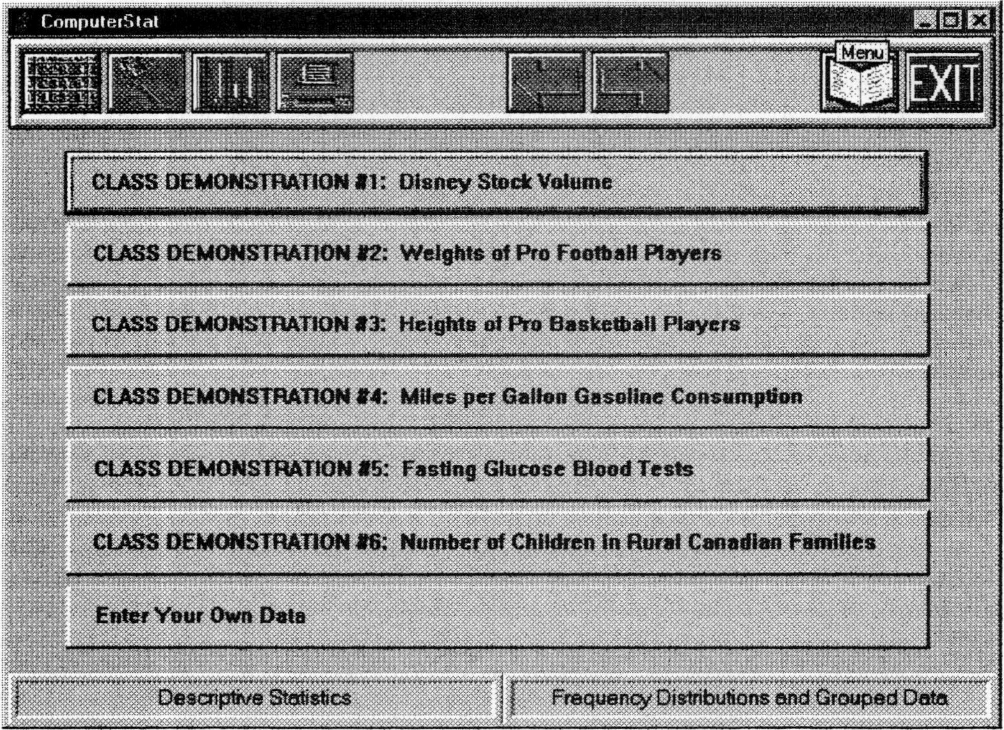

Select the class demonstration

Number of classes , N; N between 1 and 10

Option II: Enter your own data.

Number of digits to the right of the decimal point; this value can be a number from 0 through 4. Scan all your data and find the value with the most places after the decimal point. This is the number you will need to tell the computer. If you have more than 4 digits after the decimal point, you will either need to round your data or scale it before entering it into the program.

Number of pieces of data, M; M between 2 and 200

Number of classes, N; N between 1 and 10

Ith data value X(I) if you enter your own data; If you make an error when entering the data, jot down the value of I where the error occurred. Then you can correct the data item.

Data correction option; If you know the Ith position in which the data error occurred, you can correct it with this option.

Categorization of data type, sample or population; This input in necessary in the options to compute the mean and standard deviation

Program Options:

1) Rerun program with same data and new classes
2) Rerun program with new data
3) Graph histogram
4) Graph ogive
5) Compute mean and standard deviation of data

Output:

List of data values
Smallest data value
Largest data value
Range
Class width
Frequency table: included are class boundaries, class frequencies, relative frequencies, class midpoints, and cumulative frequencies
Histogram
Ogive
Mean and standard deviation computed from from raw data
Mean and standard deviation estimated from the data as it appears in the frequency table (relates to Section 3.3 of *Understandable Statistics*).

II. Sample Run

Throughout the day from 8AM to 11PM Tiffany counted the number of ads occurring every hour on one commercial T.V. station. The 15 pieces of data are

10	12	8	7	15	6	5	8
8	10	11	13	15	8	9	

Use the *ComputerStat* program FREQUENCY DISTRIBUTION AND GROUPED DATA to find the smallest and largest pieces of data and the range. Using 4 classes, find the class width, make a frequency table, and graph the histogram.

First enter the data. Select Enter Your Own Data from the data menu.

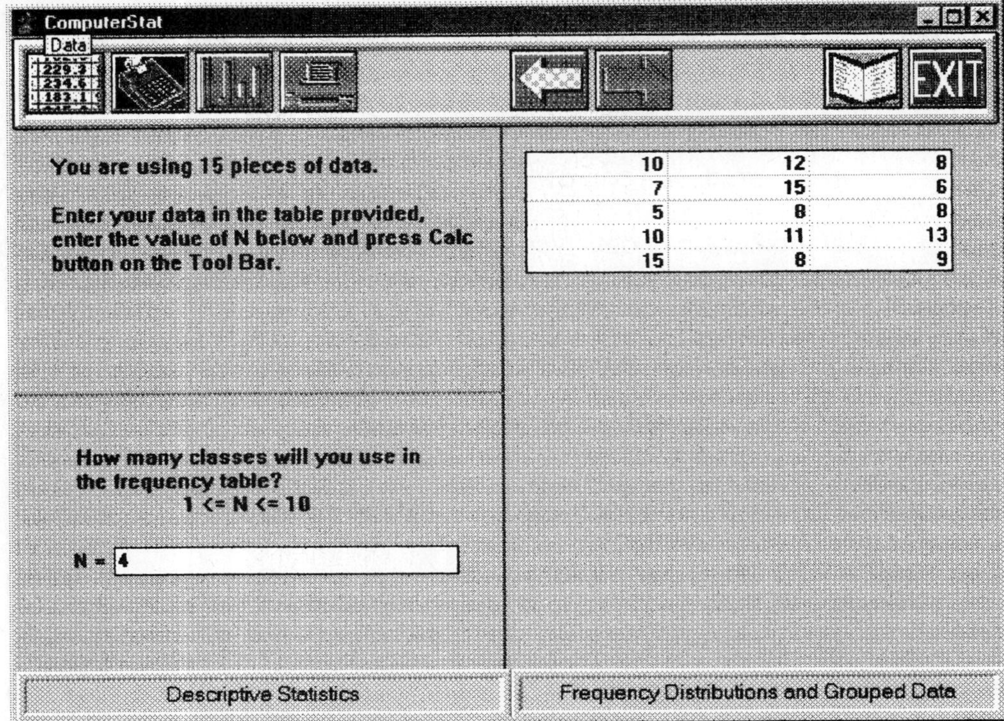

Press Calculate (the calculator)

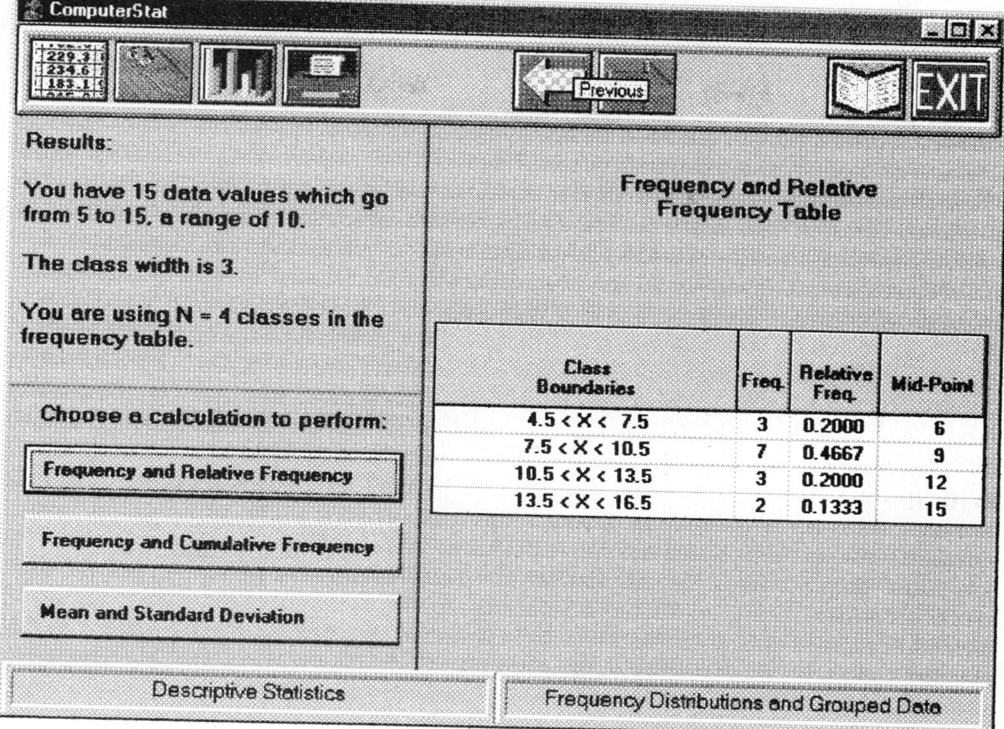

Press the Graph button for graphing options

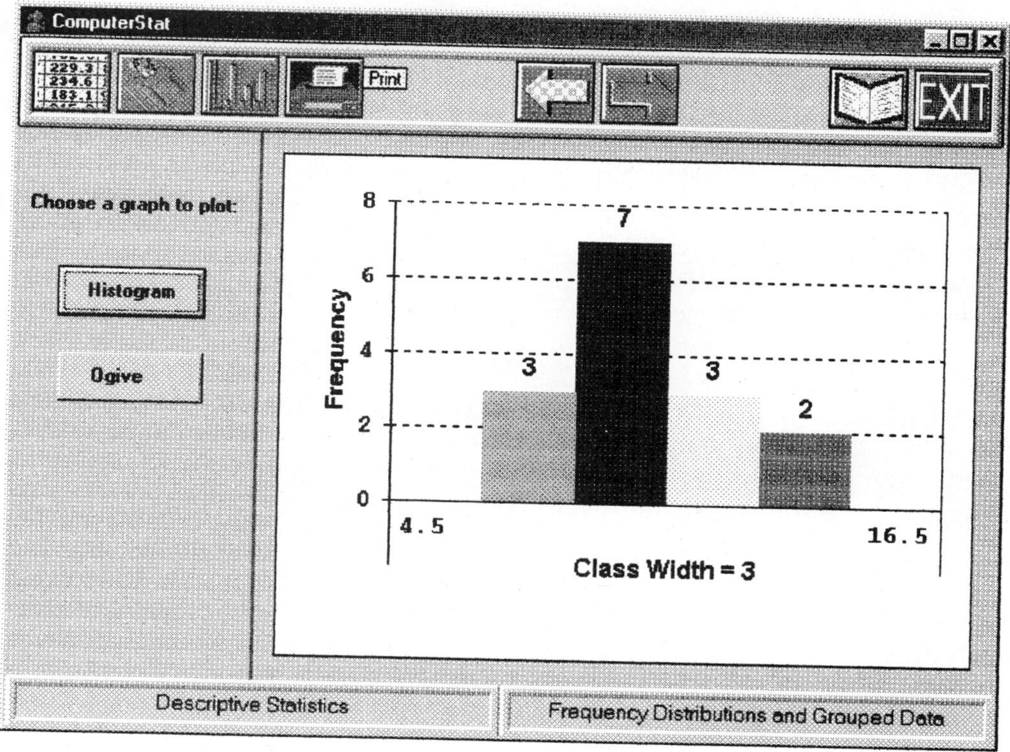

Lab Activities for FREQUENCY DISTRIBUTIONS

These activities coordinate with Section 2.3 Histograms and Frequency Distributions of *Understandable Statistics*

1. The following data represent the number of shares of Disney stock (in hundreds of shares) sold for a random sample of 60 trading days in 1993 and 1994

   SOURCE: DOW-JONES INFORMATION RETRIEVAL SERVICE

12584.00	9441.00	18960.00	21480.00	10766.00
13059.00	8589.00	4965.00	4803.00	7240.00
10906.00	8561.00	6389.00	14372.00	18149.00
6309.00	13051.00	12754.00	10860.00	9574.00
19110.00	29585.00	21122.00	14522.00	17330.00
18119.00	10902.00	29158.00	16065.00	10376.00
10999.00	17950.00	1 5418.00	12618.00	16561.00
8022.00	9567.00	9045.00	8172.00	13708.00
11259.00	10518.00	9301.00	5197.00	11259.00
10518.00	9301.00	5197.00	6758.00	7304.00
7628.00	14265.00	13054.00	15336.00	14682.00
27804.00	16022.00	24009.00	32613.00	19111.00

   Use the *CompuerStat* program FREQUENCY DISTRIBUTIONS AND GROUPED DATA. Select Class Demonstration #1 DISNEY STOCK VOLUME. Use the program to make a frequency table showing class boundaries, frequencies, relative frequencies, midpoints, and cumulative frequencies for 5 classes. Repeat the process for 9 classes. Is the data distribution skewed or symmetric? Is the distribution shape more pronounced with 5 classes or with 9 classes? Look at the ogive for 9 classes. Approximate the volume level traded on 75% of the days.

2. Explore the other class demonstration data files

   CLASS DEMONSTRATION #2: WEIGHTS OF PRO FOOTBALL PLAYERS
   CLASS DEMONSTRATION #3: HEIGHTS OF PRO BASKETBALL PLAYERS
   CLASS DEMONSTRATION #4: MILES PER GALLON GASOLINE CONSUMPTION
   CLASS DEMONSTRATION #5: FASTING GLUCOSE BLOOD TESTS
   CLASS DEMONSTRATION #6: NUMBER OF CHILDREN IN RURAL CANADA FAMILIES

   in the options menu for FREQUENCY DISTRIBUTIONS
   In each case:

Lab Activities for FREQUENCY DISTRIBUTIONS AND GROUPED DATA continued

a) Make histograms with 5, 8, and 10 of classes.
b) Categorize the shape of the distribution as: uniform, symmetric, skewed, or bimodal
c) Look at ogives when you use 10 classes. Approximately what data value do about 75% of the data fall below?

3. a) Consider the data

1	3	7	8	10
6	5	4	2	1
9	3	4	5	2

Use the *ComputerStat* program FREQUENCY DISTRIBUTIONS AND GROUPED DATA to make a frequency table and histogram with N = 3 classes. Jot down the results so that you can compare them to part (b)

b) Now add 20 to each data value of part (a). The results are

21	23	27	28	30
26	25	24	22	21
29	23	24	25	22

Make a frequency table with 3 classes. Compare the class boundaries, midoints, frequencies, relative frequencies, and cumulative frequencies with those obtained in part (a). Are each of the boundaries and midoints 20 more than those of part (a)? How do the frequencies, relative frequencies, and cumulative frequencies compare?

c) Use your discoveries from part (b) to predict the frequency table and histogram with 3 classes for the data values below.

1001	1003	1007	1008	1010
1006	1005	1004	1002	1001
1009	1003	1004	1005	1002

Would it be safe to say that we simply shift the histogram of part (a) 1000 units to the right?

d) What if we multiply each of the values of part (a) by 10? Will we effectively multiply the entries for class boundaries by 10? To explore this relation, use FREQUENCY DISTRIBUTIONS to generate the histogram and frequency table with 3 classes for the data.

Lab Activities for FREQUENCY DISTRIBUTIONS AND GROUPED DATA continued

10	30	70	80	100
60	50	40	20	10
90	30	40	50	20

Compare the frequency table of these data to the one from part (a). You will see that there does not seem to be any relation. To see why, look at the class width and compare it to the class width of part (a). The class width is always increased to the next integer value no matter how large the integer data values are. Consequently, the class width for the data in part (d) was increased to 31 instead of to 30.

4. Histograms are not effective displays for some data. Consider the data

1	2	3	6	5	7
9	8	4	12	11	15
14	12	6	2	1	206

Use FREQUENCY DISTRIBUTIONS to create frequency tables and histograms with 2 classes. Then change to 3 classes, on up to 10 classes. Notice that all the histograms lump the first 14 data into the first class, and the one data value 206 in the last class. What feature of the data causes this phenomenon? Recall that

Class width = (largest data - smallest data)/(number of classes)
increased to the next integer

How many classes would you need before you began to see the first 15 data values distributed among several classes? What would happen if you simply did not include the extreme value 206 in your histogram?

CHAPTER 3 AVERAGES AND VARIATION

AVERAGES, VARIATION, BOX-AND-WHISKER PLOTS

Sections in *Understandable Statistics*
- **3.1 Mode, Median, and Mean**
- **3.2 Measures of Variation,**
- **3.4 Percentiles and Box-and-Whisker Plots**

Main menu selection: DESCRIPTIVE STATISTICS

Sub-menu selection: AVERAGES, VARIATION, BOX-AND-WHISKER PLOTS

I. Description of the program

This program computes averages and measures of variation for raw data. The averages include median, mode, and mean. Measures of variation include the range, standard deviation and variance computed for data forming a sample or population. The program also produces the five number summary used to make a box-and-whisker plot and displays the graph of a box-and-whisker plot.

Input: Option I: Use an existing class demonstration data file that is included in the *ComputerStat* data menu. The data values included in the demonstrations are displayed in Appendix I of this guide.

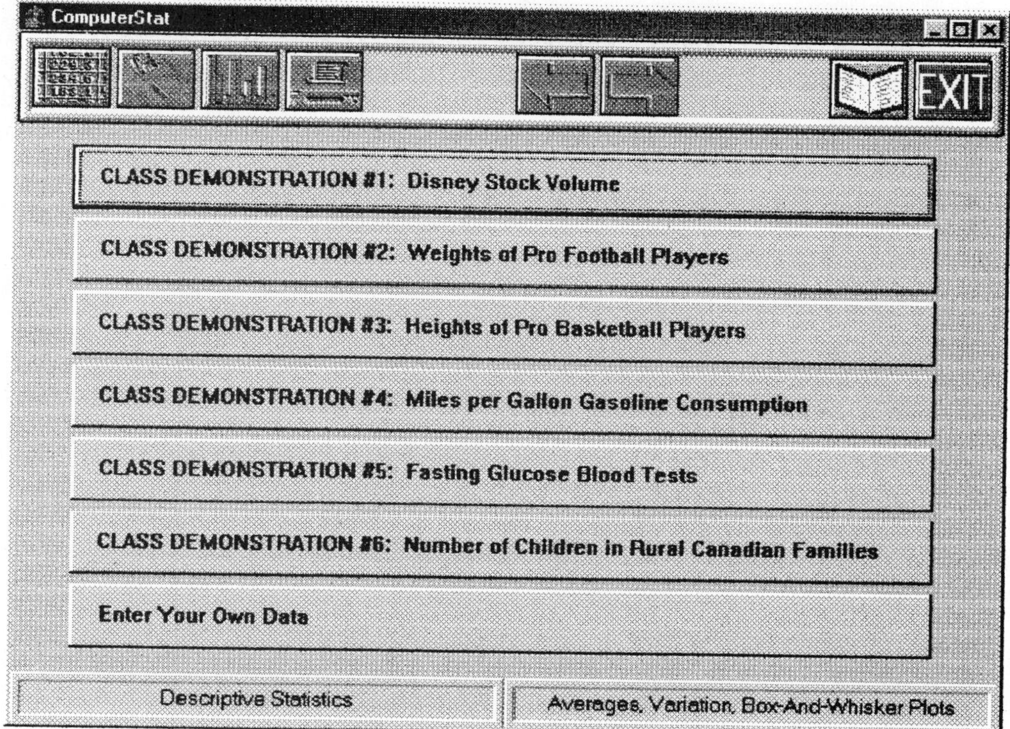

Option 2: Enter your own data

Number of pieces of data N; N between 2 and 100

Data type; Forming a sample or forming a population

Ith data value, X(I); Again by noting the Ith position of any data entry error, you will be able to correct the data as you did in the program **FREQUENCY DISTRIBUTION AND GROUPED DATA**.

Data correction option; If you know the Ith position in which the data error occurred, you can correct it with this option.

Output: List of data ordered from smallest to largest
Mode
Mean
Range
Variance; sample or population
Standard deviation and CV; sample or population
Five number summary for box-and-whisker plot
- low value
- first quartile value
- median
- third quartile value
- high value

Box-and-whisker plot

II. Sample Run

Consider the weights of Pro football players. The weights of a random sample of 50 linebackers from professional teams are included in Class Demonstration #2. Use the program to generate a statistical summary of the weights and to draw a box-and-whisker plot.

Run the program, and select

2) CLASS DEMONSTRATION #2: WEIGHTS OF PRO FOOTBALL PLAYERS
   You will see a data screen listing the data. When you click on the calculate button (calculator key) this screen appears. Click on Mode, Mean, Median to get the left side.

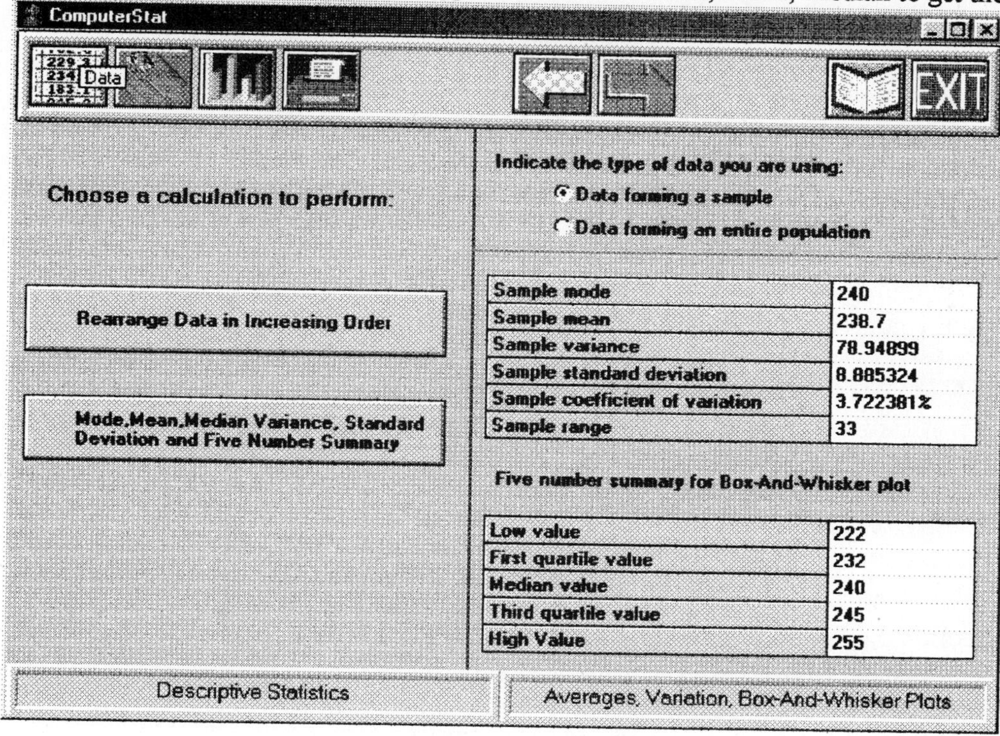

Click on the graph button to get a box-and-whisker plot

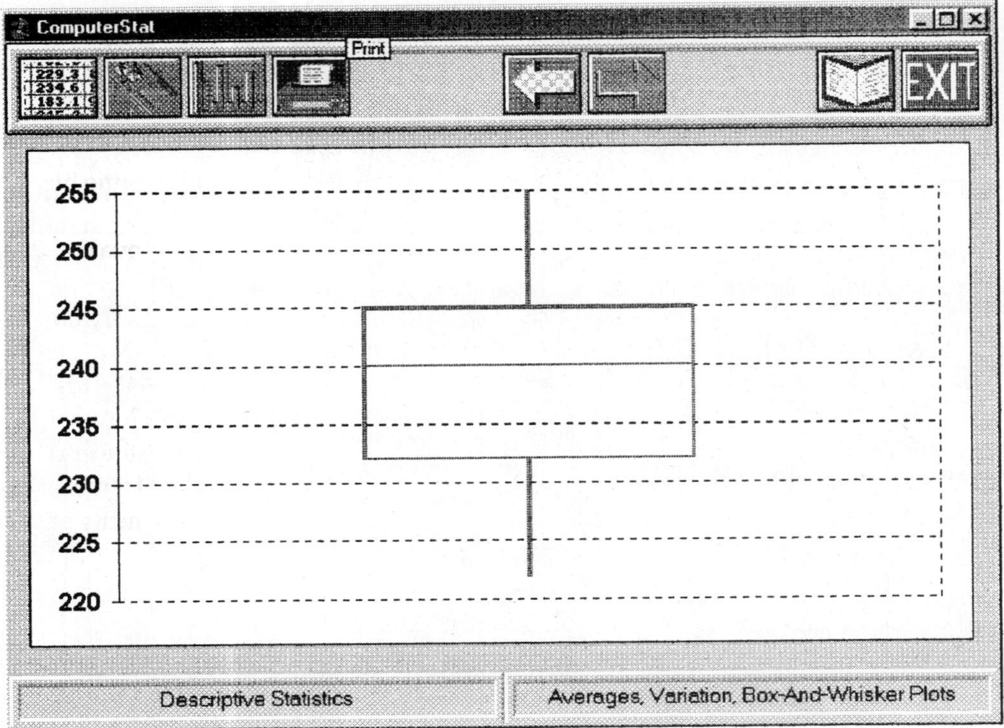

Lab Activities for AVERAGES, VARIATION, BOX-AND-WHISKER PLOTS

These activities coordinate with the following sections in *Understandable Statistics*
Section 3.1 Mode, Median, and Mean
Section 3.2 Measures of Variation
Section 3.4 Percentiles and Box-and-Whisker Plots.

1. Select one of the class demonstrations available in the options menu

<<<<< OPTIONS AVAILABLE >>>>>

1) CLASS DEMONSTRATION #1: DISNEY STOCK VOLUME
2) CLASS DEMONSTRATION #2: WEIGHTS OF PRO FOOTBALL PLAYERS
3) CLASS DEMONSTRATION #3: HEIGHTS OF PRO BASKETBALL PLAYERS
4) CLASS DEMONSTRATION #4: MILES PER GALLON GASOLINE CONSUMPTION.
5) CLASS DEMONSTRATION #5: FASTING GLUCOSE BLOOD TESTS
6) CLASS DEMONSTRATION #6: NUMBER OF CHILDREN IN RURAL CANADIAN FAMILIES

Lab Activities for AVERAGES, VARIATION, BOX-AND-WHISKER PLOTS continued

Use the program AVERAGES, VARIATION, BOX-AND-WHISKER PLOTS to generate a statistical summary of the data. Then, use the program FREQUENCY DISTRIBUTIONS AND GROUPED DATA with the same class demonstration to generate a histogram and ogive with 7 classes.

Use the information from both programs to answer the following questions:

a) Is the distribution skewed or symmetric? How is this shown in both the histogram and the box-and-whisker plot?

b) Look at the box-and-whisker plot. Are the data more spread out above the median or below the median?

c) Look at the histogram and estimate the location of the mean on the horizontal axis. Are the data more spread out above the mean or below the mean?

d) Do there seem to be any data values that are unusually high or unusually low? If so, how do these show up on a histogram or on a box-and-whisker plot?

e) Pretend that you are writing a brief article for a newspaper. Describe the information about the data in the class demonstration you selected in non technical terms. Be sure to make some comments about the "average" of the data measurements and some comments about the spread of the data.

2. a) Consider the test scores of 30 students in a political science class.

85	73	43	86	73	59	73	84	62	100
75	87	70	84	97	62	76	89	90	83
70	65	77	90	84	80	68	91	67	79

For this population of test scores find the mode, median, mean, range, variance, standard deviation, CV, and the five number summary and make a box-and-whisker plot. Be sure to record all of these values so you can compare them to the results of part (b).

b) Suppose Greg was in the political science class of part (a). Suppose he missed a number of classes because of illness, but took the exam anyway and made a score of 30 instead of 85 as listed as the first entry of the data in part (a). Again, use the program to find the mode, median, mean, range, variance, standard deviation, CV, and the five number summary and make a box-and-whisker plot using the new data set. Compare these results to the corresponding results of part (a). Which average was most affected: mode, median, or mean? What about the range, standard deviation, and coefficient of variation? How do the box-and-whisker plots compare?

c) Write a brief essay in which you use the results of parts (a) and (b) to predict how an extreme data value affects a data distribution. What do you predict for the results if Greg's test score had been 80 instead of 30 or 85?

Lab Activities for AVERAGES, VARIATION, BOX-AND-WHISKER PLOTS continued

3. In this problem we will explore the effects of changing data values by multiplying each data value by a constant, or by adding the same constant to each data value.

   a) Consider the data

1	8	3	5	7
2	10	9	4	6
3	5	2	9	1

   Use the computer program to find the mode (if it exists), mean, sample standard deviation, range, and median and box-and-whisker plot. Make a note of these values since you will compare them to those obtained in parts (b) and (c).

   b) Now multiply each data value of part (a) by 10 to obtain the data

10	80	30	50	70
20	100	90	40	60
30	50	20	90	10

   Again use the computer program to find the mode (if it exists), mean, sample standard deviation, range, and median and box-and-whisker plot. Compare these results to the corresponding ones of part (a). Which values changed? Did those that changed change by a factor of 10? Did the range or standard deviation change? Referring to the formulas for these measures (see Section
   3.2 of *Understandable Statistics*) can you explain why these values behaved the way they did? Will these results generalize to the situation of multiplying each data entry by 12 instead of by 10? What about multiplying each by 0.5? Predict the corresponding values that would occur if we multiplied the data set of part (a) by 1000.

   c) Now suppose we add 30 to each data value of part (a)

32	38	33	35	37
32	40	39	34	36
33	35	32	39	31

   Again use the computer program to find the mode (if it exists), mean, sample standard deviation, range, and median and box-and-whisker plot. Compare these results to the corresponding ones of part (a). Which values changed? Of those that are different, did each change by being 30 more than the corresponding value of part (a)? Again look at the formulas for range and standard deviation. Can you predict the observed behavior from the formulas? Can you generalize these results? What is we added 50 to each data value of part (a). Predict the values for the mode, mean, median, sample standard deviation and range.

CHAPTER 4 ELEMENTARY PROBABILITY THEORY

Simulation of Probability Experiments

(Sections 4.1 and 4.2 of *Understandable Statistics*)

Main Menu Selection: DESCRIPTIVE STATISTICS

Sub-menu selection: RANDOM SAMPLES

There are three programs within this menu option

Select a Random Sample from a Population
Simulate the Experiment of Tossing One Die
Simulate the Experiment of Tossing Two Dice

In Chapter 2, the program Select a Random Sample from a Population was discussed.

Description of the program Simulate the Experiment of Tossing one Die

Input: The number of times you want to toss the die (from 1-1,000,000 times)

Output: The number of each of the outcomes 1 through 6, and the relative frequency of each outcome.

Sample Run: Simulate tossing a die 400 times.

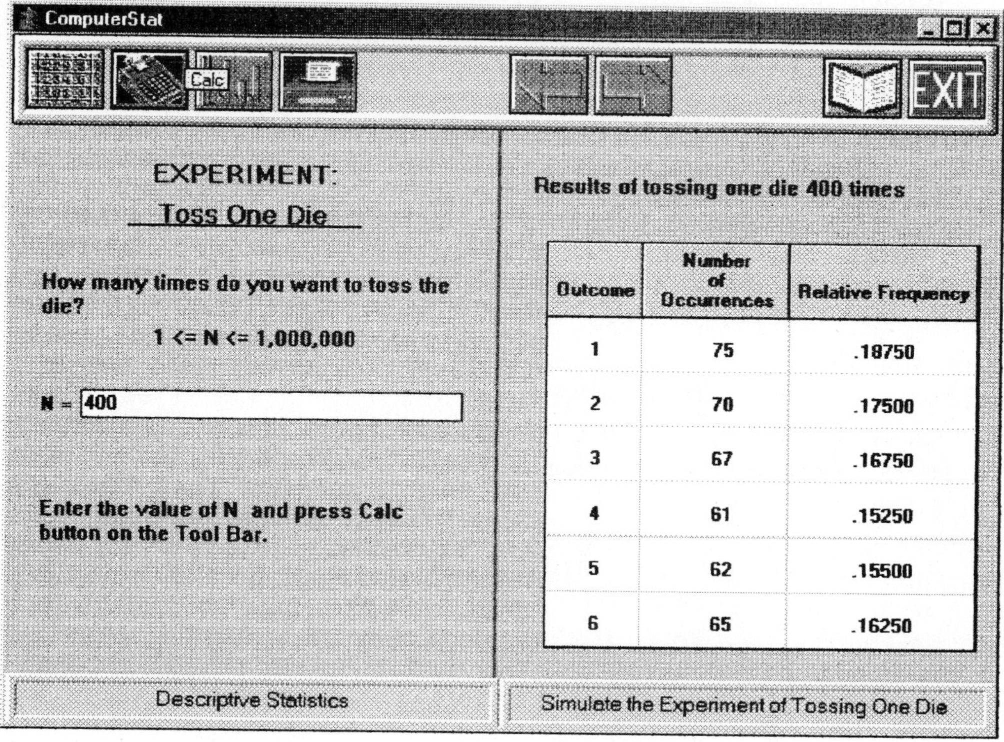
ComputerStat
Calc
EXIT
EXPERIMENT:
Toss One Die
How many times do you want to toss the die?
1 <= N <= 1,000,000
N = 400
Enter the value of N and press Calc button on the Tool Bar.
Results of tossing one die 400 times
Descriptive Statistics
Simulate the Experiment of Tossing One Die

Outcome	Number of Occurrences	Relative Frequency
1	75	.18750
2	70	.17500
3	67	.16750
4	61	.15250
5	62	.15500
6	65	.16250

Lab Activities for Probability

These activities coordinate with Sections 4.1 and 4.2 of *Understandable Statistics*

1. In the program RANDOM SAMPLES select the option to simulate the experiment of tossing one die. Since the possible outcomes 1, 2, 3, 4, 5, 6 are equally likely, the theoretical probability for each outcome is 1/6 or approximately 0.16667.

   a) Choose N = 6 tosses. Did you get an outcome of 2? How many times? Repeat the experiment again with N = 6. Did you get an outcome of 2 this time? Did it occur as many times as before? Repeat the experiment a third time. Should you expect the same results as you got before? Does the fact that the theoretical probability P(outcome is 2) = 1/6 guarantee that every time you toss a die six times you will get exactly one outcome of 2? Why or why not?

   b) Repeat the experiment of using number of tosses N = 50, 100, 500, 1000, and 3000. Record the relative frequency for the outcome of 2. How do the relative frequencies compare to the theoretical results P(outcome is 2) = 0.1667? Does it appear that more trials lead to results that match the theory better?

   c) When you toss a die, the outcomes are mutually exclusive since it is impossible to show two distinct numbers on the top of one die. Therefore, theoretically,
   P(even outcome) = P(2 *or* 4 *or* 6) = P(2) + P(4) + P(6) = 0.5

   Run the option to toss one die. Let the number of tosses be 5000. Using the number of occurrences of an outcome, compute the probability of obtaining a 2 or 4 or 6. In other words, add up the total occurrences of even numbers and divide by the notal number of trials, 5000. How does this result compare to the theoretical result of 0.5? Do you get the same experimental result if you add the relative frequencies of obtaining a 2 or 4 or 6? Does the addition rule seem to work for mutually exclusive events?

2. In the program RANDOM SAMPLES select the option to simulate the experiment of tossing two dice. Use N = 1000 tosses.
   a) Use relative frequencies to estimate P(2 on 1st die *and* 5 on second). How does this result compare to the theoretically computed result using the multiplication rule for independent events?
   P(2 on 1st *and* 5 on second) = P(2 on 1st)·P(2 on 2nd) = 1/36 or about 0.02778
   b) Repeat part (a) for 5000 tosses. Do the experimental and theoretical results agree better?
   c) Repeat parts (a) and (b) for P(6 on 1st die *and* odd on second).

Lab Activities for Probability continued

3. In the program RANDOM SAMPLES, the option to simulate the experiment of tossing two dice also gives the relative frequencies of the possible sums. Let the number of tosses be N = 1000.
   a) Use one of the screens showing the 36 possible outcomes possible when you toss two dice once to estimate the theoretical probabilities

      P(sum of 8); P(sum of 9); P(sum of 10); P(sum of 11); P(sum of 12)

      Compare these results with the theoretical probabilities given earlier in this chapter in the problem used in the Sample run.

   b) Continue with the program until you see the screen showing the possible sums and their relative frequencies. How do these numbers compare to the results you got in part (a)?

   c) Since the events are mutually exclusive, we can compute

      P(sum more than 10) = P(sum of 11) · P(sum of 12)

      Using the results of part (a), estimate the theoretical probability of a sum more than 10.

   d) Using the computer output, estimate the probability of getting a sum of more than 10 by adding the number of occurrences of sums more than 10 and dividing by the number of times the dice were tossed, 1000. How does this result compare to the estimated probability of part (c)?

   e) Repeat parts (a) through (d) using n = 5000 trials.

4. Assume that the outcomes of having a boy or girl baby are equally likely. We can experimentally estimate the probability that a family with three children has all girls. In the program RANDOM SAMPLES, select the option of choosing a sample from a population. Assign the outcome 1 to boy and 2 to girl, so our "population" consists of the two numbers 1 and 2 and N = 2. Sample with replacement since the outcomes boy and girl are independent. Choose a sample size = 3 to reflect that we are looking at families with three children.

   a) Record the outcomes in the sample. The list 1 2 2 translates to boy, girl, girl.

   b) Repeat the experiment for 29 more families of three children and record the outcomes of each of the families. Each time use population size N = 2, sample with replacement, and use sample size M = 3.

Lab Activities for Probability continued

c) Now you have a total of 30 groups of 3 children. How many of these groups contain all girls? What is the experimental estimate for the probability that a family of three children consists of there girls? In Chapter 5 of *Understandable Statistics* you will see how to compute the theoretical probability of having three girls out of three children and we ask you to do this in the Lab Activities for Binomial Probabilities.

CHAPTER 5 THE BINOMIAL PROBABILITY DISTRIBUTION AND RELATED TOPICS

EXPECTED VALUE AND STANDARD DEVIATION OF DISCRETE PROBABILITY DISTRIBUTIONS

(Section 5.1 of *Understandable Statistics*)

Main menu selection: PROBABILITY DISTRIBUTIONS AND CENTRAL LIMIT THEOREM

Sub-menu selection: EXPECTED VALUE AND STANDARD DEVIATION OF DISCRETE PROBABILITY DISTRIBUTIONS

I. Description of the program

The program computes the expected value (mean), variation, and standard deviation of a discrete probability distribution. The user enters the discrete random variables X and corresponding probabilities P(X). The X values cannot be repeated and the probability values P(X) must be between 0 and 1 inclusive and must add up to 1 within round off error of 0.002.

Input: Number of X values N; N between 1 and 50
Discrete random variables; X
Corresponding probabilities, P(X); P(X) between 0 and 1
At the sound of the beep, check your data and record the entry number of any error. For instance, if the error occurred at the data entry 7 out of 10, note the value 7 as the entry #.
Data correction option; If you made an error, use this option with the entry number t o correct X or P(X)..

II. Sample run

If two fair dice are tossed, the probability for the possible sums shown on the dice are

Sum X	2	3	4	5	6	7	8	9	10	11	12
P(X)	.028	.056	.083	.111	.139	.166	.139	.111	.083	.056	.028

Find the expected sum if you roll two dice. Find the variance and standard deviation of the X distribution.

Select the program. Expected Value and Standard Deviation of Discrete Probability Distributions.

Next, enter the value 11 for N since we have 11 different possible sums when we toss two dice. Click on OK. Begin entering your data.

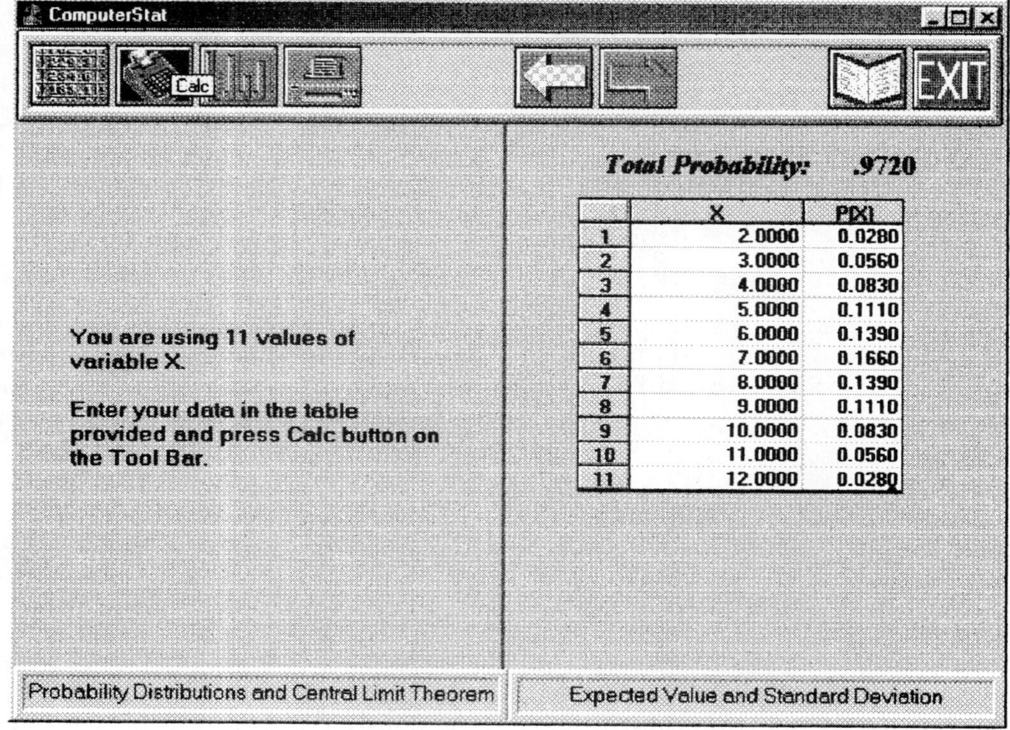

	X	P[X]
1	2.0000	0.0280
2	3.0000	0.0560
3	4.0000	0.0830
4	5.0000	0.1110
5	6.0000	0.1390
6	7.0000	0.1660
7	8.0000	0.1390
8	9.0000	0.1110
9	10.0000	0.0830
10	11.0000	0.0560
11	12.0000	0.0280

Click on Calc.

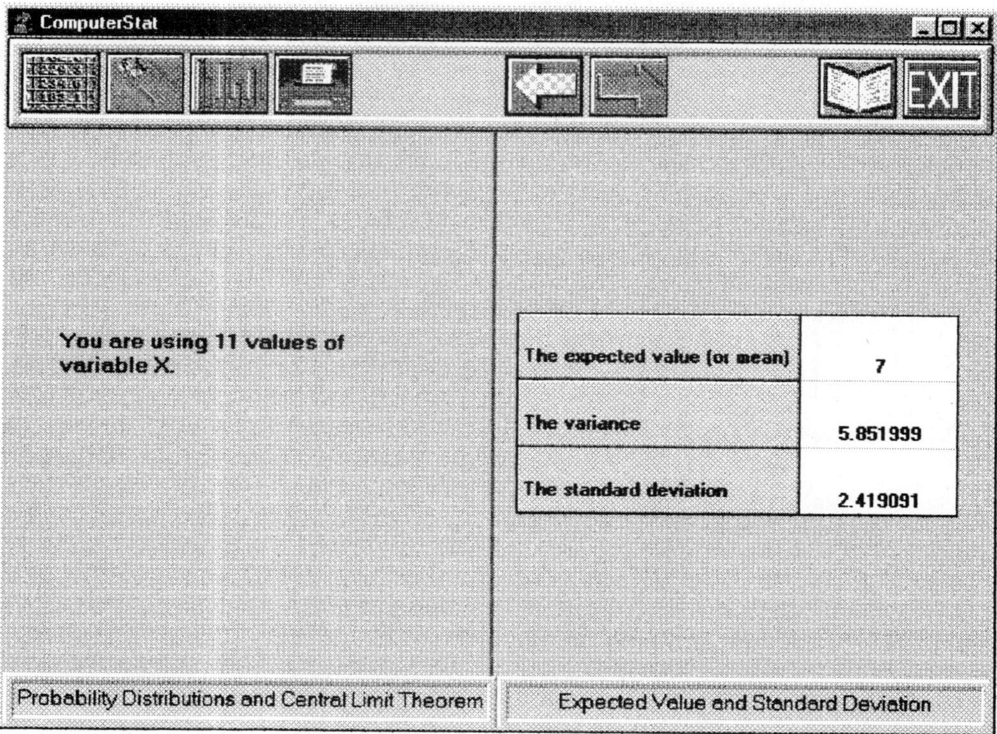

The expected value (or mean)	7
The variance	5.851999
The standard deviation	2.419091

BINOMIAL COEFFICIENTS AND PROBABILITY DISTRIBUTION
(Sections 5.2, 5.3, of *Understandable Statistics)*

Main menu selection: PROBABILITY DISTRIBUTIONS AND CENTRAL LIMIT THEOREM

Sub-menu selection: BINOMIAL COEFFICIENTS AND PROBABILITY DISTRIBUTION

I. Description of program

Each trial in a binomial experiment has two possible outcomes: success or failure In this program the user inputs the number N of binomial trials and the probability of success on any one trial. The program displays a table of binomial coefficients C and binomial probabilities P(R) where R is the number of successes in N trials,. The cumulative probability P(0 <= X <= R) is also displayed. Note that R can have integer values from 0 to N inclusive. There is an option to graph the distribution.

Input: The number of trials, N; N between 1 and 40 inclusive
Probability of success on a single trial, P; P between 0 and 1 inclusive

Output: A table of binomial coefficients C and probabilities P(R) and cumulative probabilities P(0 <= X <= R) for R success out of N trials. Also a graph and values of μ and σ.

II. Sample Run

A surgeon performs a difficult spinal column operation. The probability of success of the operation is P = 0.73. Ten such operations are schedules. Find the probability of success for 0 through 10 successes out of these 10 operations.

Select the program BINOMIAL COEFFICIENTS AND PROBABILITY DISTRIBUTION. The first screen shows the program title and gives a brief description. Use the number of trials N = 10, and the probability of success P = 0.73. Click on the calculate button.

The outcome is the following table.

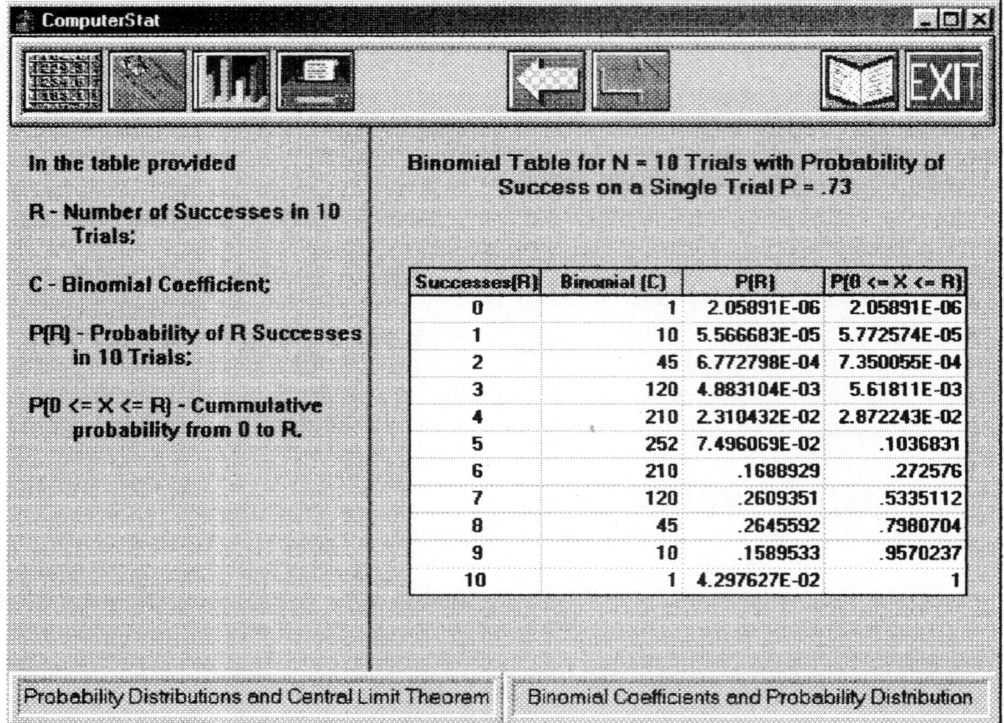

ComputerStat

In the table provided

R - Number of Successes in 10 Trials;

C - Binomial Coefficient;

P(R) - Probability of R Successes in 10 Trials;

P(0 <= X <= R) - Cummulative probability from 0 to R.

Binomial Table for N = 10 Trials with Probability of Success on a Single Trial P = .73

Successes(R)	Binomial (C)	P(R)	P(0 <= X <= R)
0	1	2.05891E-06	2.05891E-06
1	10	5.566683E-05	5.772574E-05
2	45	6.772798E-04	7.350055E-04
3	120	4.883104E-03	5.61811E-03
4	210	2.310432E-02	2.872243E-02
5	252	7.496069E-02	.1036831
6	210	.1688929	.272576
7	120	.2609351	.5335112
8	45	.2645592	.7980704
9	10	.1589533	.9570237
10	1	4.297627E-02	1

Probability Distributions and Central Limit Theorem | Binomial Coefficients and Probability Distribution

To graph the distribution, click on the graph button.

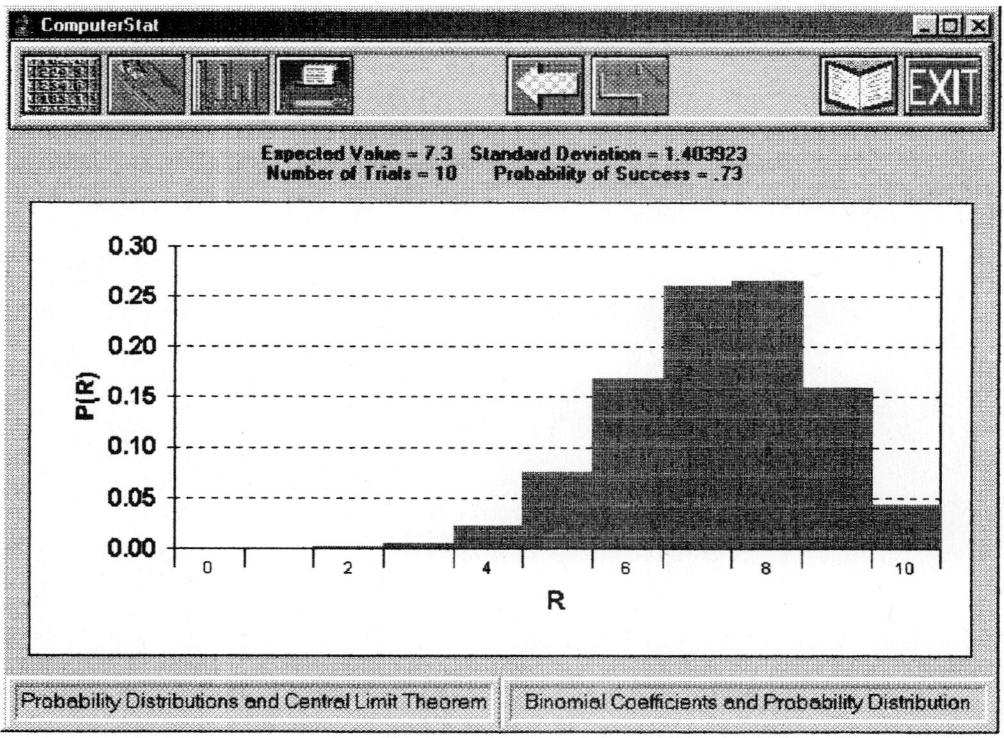

Lab Activities for Discrete Probability Distributions

These activities coordinate with Section 5.1 of *Understandable Statistics*

1. Hold Insurance has calculated the following probabilities for claims on a new $15,000 vehicle for one year if the vehicle is driven by a single male in his twenties.

$Claim, X:	0	1000	5000	10000	15000
P(X)	0.63	0.24	0.10	0.02	0.01

Find the expected value of a claim in one year. What is the standard deviation of the claim distribution? What should the annual premium be to include $200 in overhead and profit as well as the expected claim?

2. Quick Dental Services claims that the probability distribution for waiting times between the actual appointment time and the time the patient sees the dentists is (in minutes)

Waiting time, X:	5	10	15	20	25	30	35	40	45
P(X) :	0.43	0.28	0.11	0.05	0.04	0.03	0.03	0.02	0.01

Find the expected waiting time and the standard deviation of the waiting time distribution. Based on this information, what would be a reasonable waiting time to advertise?

3. Suppose the possible outcomes of an experiment are labeled 1 through 5, and the reported probabilities of each outcome are as follows

X:	1	2	3	4	5
P(X):	0.30	0.25	0.15	0.10	0.15

Enter these values in the program. Why is the probability distribution given above not correct? Do you suppose there is another unreported outcome?

Lab Activities using BINOMIAL COEFFICIENTS AND PROBABILITY DISTRIBUTION

These activities relate to Sections 5.2, 5.3 in *Understandable Statistics*

1. You toss a coin 8 times. Call heads success. If the coin is fair, the probability of success P is 0.5. What is the probability of getting exactly 5 heads out of 8 tosses? of getting exactly 20 heads out of 40 tosses? of getting 20 heads or fewer out of 40 tosses? of getting more than 20 heads out of 40 (Hint: Subtract the probability of 20 heads or fewer from 1)?

2. At Pleasant College, the student services office has determined that the probability a student will get the class schedule he or she wants is 0.83. Five friends register. What is the probability that all five of them get their preferred schedule? What is the probability that at least 3 of them get their preferred schedule? Next graph the probability distribution. Is skewed or symmetric? Note the expected value and standard deviation shown above the graph. What is the expected number out of 5 students to get their requested schedule?

3. Some tables for the binomial distribution give values only up to 0.5 for the probability of success p. There is a symmetry to the values for p greater than 0.5 with those values of p less than 0.5.

   a) Examine the binomial distribution table in *Understandable Statistics* and see if you can detect the symmetry. Compare the entries for number of successes n = 3 and probability of success p = 0.6 and p = 0.4.

   b) Now use the *ComputerStat* program BINOMIAL COEFFICIENTS AND PROBABILITY DISTRIBUTION and graph the two distributions with N = 3 and P = 0.6 and P = 0.4. How do the graphs compare? Do the graphs show the same symmetry that the table shows?

   c) Repeat part (b) with N = 10 and p = 0.5 and p = 0.75.

   d) In general, if P! = 1 - P2, how will the graphs compare the same number of trials?

4. In Lab Activities for Probabilities, problem 7, you experimentally estimated the probability that in a family of three children, all three were girls. We can view this experiment as a binomial experiment. There are two outcomes for each child: girl which we will call success and boy which we will call failure. The probability of success for each child is 0.5. There are three trials since we are looking at three children.

   a) Return to problem 7 of Lab Activities for Probabilities and record the results of part (c). This is the experimental result obtained by using 30 families of 3 children each.

   b) Use the program BINOMIAL COEFFICIENTS AND PROBABILITY DISTRIBUTION to find the theoretical probability of 3 successes out of 3 trials when the probability of success is 0.5 and there are N = 30 trials.

   c) Graph the distribution. Above the graph you will see the mean and standard deviation. What is the expected number of girls in a family selected at random with three children?

CHAPTER 6 NORMAL DISTRIBUTIONS

CONTROL CHARTS
(Section 6.1 of *Understandable Statistics*)

Main menu selection: PROBABILITY DISTRIBUTIONS AND CENTRAL LIMIT THEOREM

Sub-menu selection: CONTROL CHARTS

I. Description of program

This program asks the user for sequential data (eg: time series, where data values are separated by equal time units). The sample mean and standard deviation are computed. Then the user is asked for a target mean and target standard deviation to be used in the construction of the control chart. A control chart is printed.

Input: Option 1: Use an existing class demonstration data file that is included in the *ComputerStat* data menu. The data values included in the demonstrations are displayed in Appendix I of this guide.

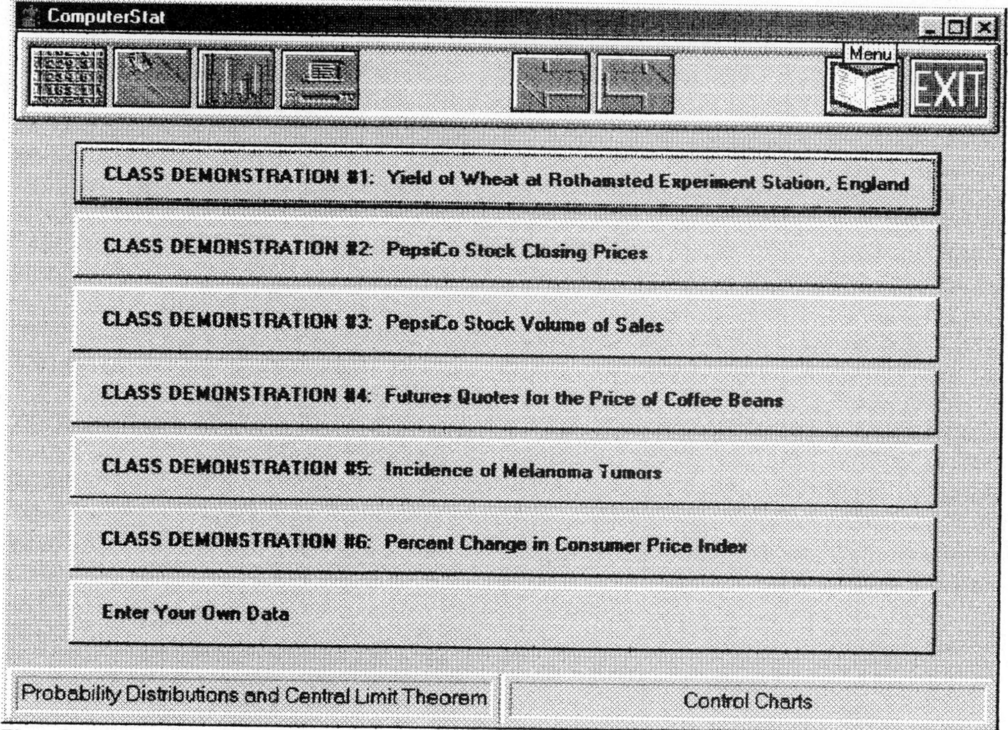

Target values for mu and sigma are provided with the data file. However, the user may change these.

Option 2: Enter your own data.
Sample size N; N between 5 and 50 inclusive
Target values for mu and sigma

Output: A display of the data including target values for mu and sigma

The control chart graph. Data values more than 3 standard deviations above the mean mu are designated by the letter A. Data values more than 3 standard deviations below the mean mu are designated by the letter B.

II. Sample Run

Let's look at a control chart for the wheat yield in an experimental plot of land at Rothamsted Experiment Station. Select the program CONTROL CHARTS, click on the data button and select Class Demonstration #1. Click on Calculate (Calculator button). Next click on Graph (the button showing a graph). The control chart follows.

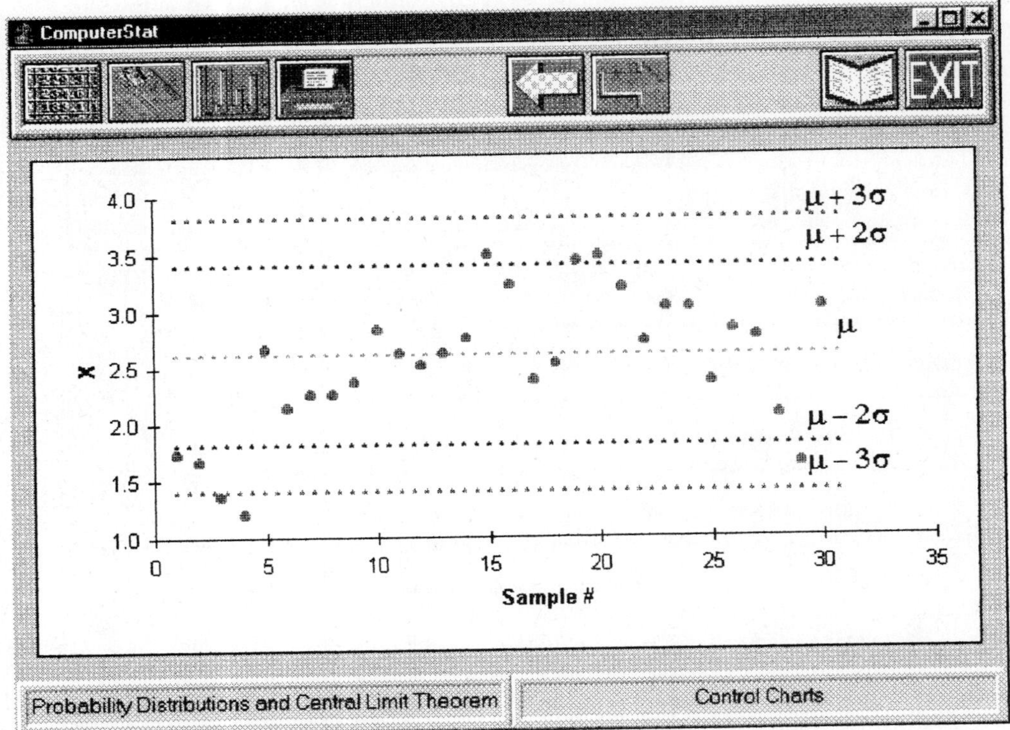

Lab Activities for CONTROL CHARTS

Use the *ComputerStat* program CONTROL CHARTS to make control charts for each of the class demonstration options included with the program. Determine if any out of control signals are present (*see* Section 6.1 of *Understandable Statistics).*

<<<<< OPTIONS AVAILABLE >>>>>

1) CLASS DEMONSTRATION #1: YIELD OF WHEAT AT ROTHAMSTED EXPERIMENT STATION, ENGLAND
2) CLASS DEMONSTRATION #2: PepsiCo STOCK CLOSING PRICES
3) CLASS DEMONSTRATION #3: PepsiCo STOCK VOLUME OF SALES
4) CLASS DEMONSTRATION #4: FUTURES QUOTES FOR THE PRICE OF COFFEE BEANS
5) CLASS DEMONSTRATION #5: INCIDENCE OF MELANOMA TUMORS
6) CLASS DEMONSTRATION #6: PERCENT CHANGE IN CONSUMER PRICE INDEX

CHAPTER 7 INTRODUCTION TO SAMPLING DISTRIBUTIONS

CENTRAL LIMIT THEOREM

(Section 7.2 of *Understandable Statistics*)

Main menu selection: PROBABILITY DISTRIBUTIONS AND CENTRAL LIMIT THEOREM

Sub-menu selection: CENTRAL LIMIT THEOREM

I. Description of Program

This program gives a demonstration of the Central Limit Theorem. The Central Limit Theorem says that if x is a random variable with any distribution having mean μ and standard deviation σ then the distribution of sample means $\bar{x}$ based on random samples of size n is such that for sufficiently large n:

a) The mean of the $\bar{x}$ distribution is approximately the same as the mean of the x distribution.

b) The standard deviation of the $\bar{x}$ distribution is approximately $\sigma/\sqrt{n}$

c) The $\bar{x}$ distribution is approximately a normal distribution

Furthermore, as the sample size n becomes larger and larger, the approximations mentions in (a), (b) and (c) become better.

We can use *ComputerStat* to demonstrate the Central Limit Theorem. The computer does not prove the theorem. A proof of the Central Limit Theorem requires advanced mathematics and is beyond the scope of an introductory course. However, we can use the computer to gain a better understanding of the theorem.

To demonstrate the Central Limit Theorem, we need a specific x distribution. One of the simplest is the uniform probability distribution. Let us compare the uniform distribution with the normal distribution.

The normal distribution is the usual bell shaped curve, but the uniform distribution is the rectangular or box shaped graph. The two distributions are very different.

The uniform distribution has the property that all subintervals of the same length inside the interval 0 to 9 have the same probability of occurrence no matter where they are located. This means that the uniform distribution on the interval from 0 to 9 could be represented on the computer by selecting random numbers from 0 to 9. Since all numbers from 0 to 9 would be equally likely to be chosen, we say we are dealing with a uniform (equally likely) probability distribution. Note that when we say we are selecting random numbers from 0 to 9, we do not just mean whole numbers or integers; we mean real numbers in decimal form such as 2.413912, and so forth.

Uniform Distribution and a Normal Distribution

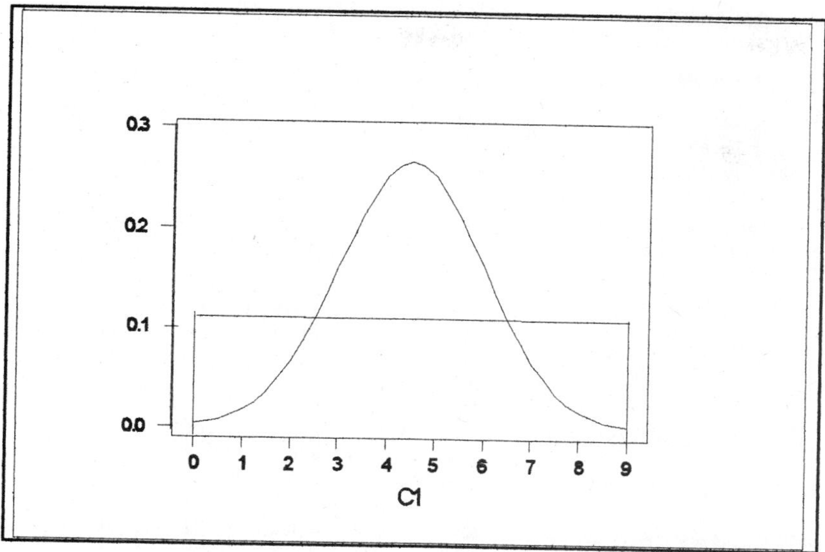

Because the interval from 0 to 9 is 9 units long and because the total area under the probability graph must be 1 (why?) then the height of the uniform probability graph must be 1/9. The mean of the uniform distribution on the interval from 0 to 9 is the balance point. Looking at the Figure, it is fairly clear that the mean is 4.5. Using advanced methods of statistics, it can be shown that for the uniform probability distribution x between 0 and 9

$$\mu = 4.5 \text{ and } \sigma = 3\sqrt{3}/2 \approx 2.598$$

The figure shows us that the uniform x distribution and the normal distribution are quite different. However, using the computer we will construct one hundred sample means $\bar{x}$ from the x distribution using a sample size of N = 5, 10, 15, 20, 30. We will see that even though the uniform distribution is very different from the normal distribution, the distribution of sample means $\bar{x}$ comes closer and closer to a bell shaped normal distribution. This surprising result is predicted by the Central Limit Theorem.

Input: Sample size for each of the 100 samples, N; N between 2 and 50 inclusive

Output: The means from the 100 samples

The sample mean of the 100 means

The sample standard deviation of the 100 means

A comparison table showing the tally of sample means falling in the specified intervals and the tally predicted by the Central Limit Theorem.

A histogram of the $\bar{x}$ distribution compared to the predicted normal curve..

II Sample Run

Run the program using sample size N = 40.

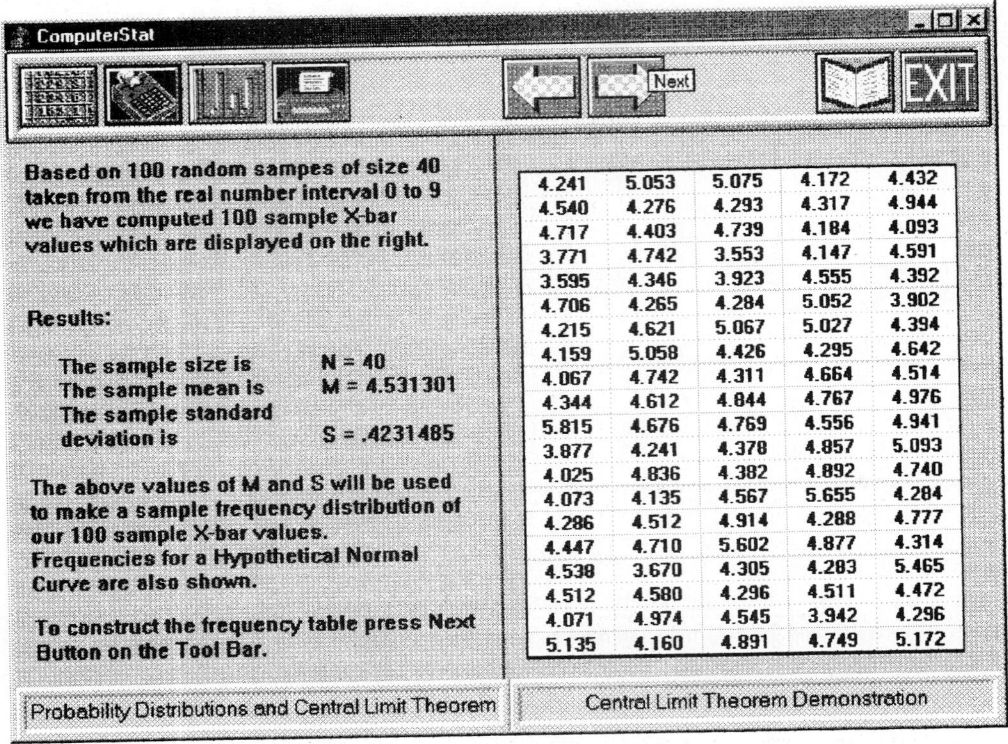

4.241	5.053	5.075	4.172	4.432
4.540	4.276	4.293	4.317	4.944
4.717	4.403	4.739	4.184	4.093
3.771	4.742	3.553	4.147	4.591
3.595	4.346	3.923	4.555	4.392
4.706	4.265	4.284	5.052	3.902
4.215	4.621	5.067	5.027	4.394
4.159	5.058	4.426	4.295	4.642
4.067	4.742	4.311	4.664	4.514
4.344	4.612	4.844	4.767	4.976
5.815	4.676	4.769	4.556	4.941
3.877	4.241	4.378	4.857	5.093
4.025	4.836	4.382	4.892	4.740
4.073	4.135	4.567	5.655	4.284
4.286	4.512	4.914	4.288	4.777
4.447	4.710	5.602	4.877	4.314
4.538	3.670	4.305	4.283	5.465
4.512	4.580	4.296	4.511	4.472
4.071	4.974	4.545	3.942	4.296
5.135	4.160	4.891	4.749	5.172

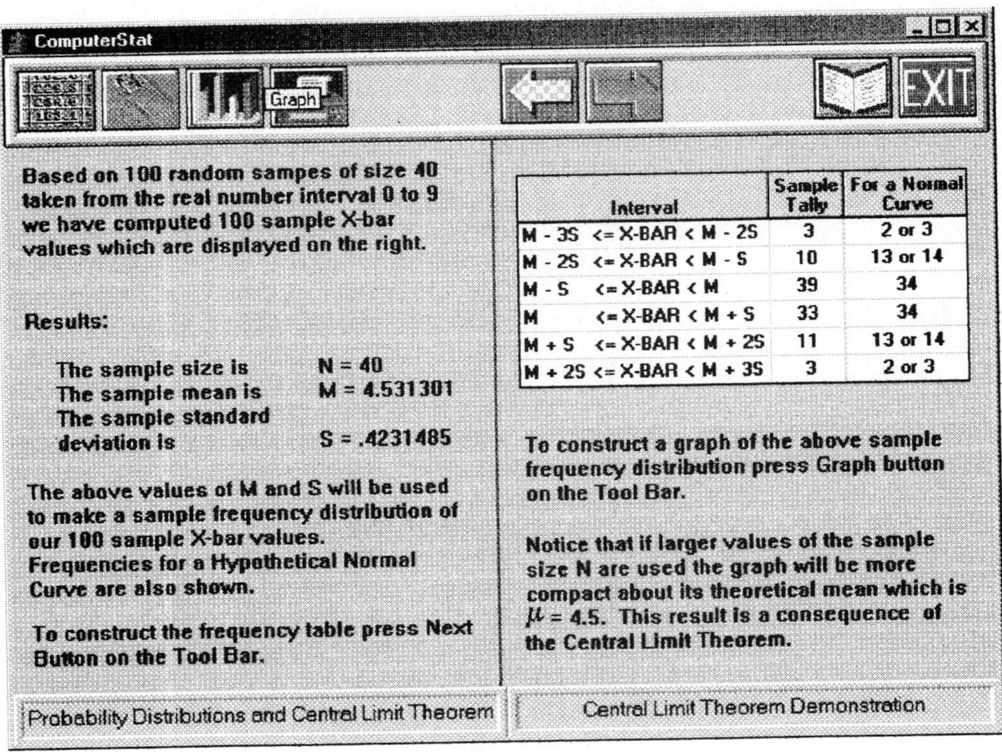

Interval	Sample Tally	For a Normal Curve
M - 3S <= X-BAR < M - 2S	3	2 or 3
M - 2S <= X-BAR < M - S	10	13 or 14
M - S <= X-BAR < M	39	34
M <= X-BAR < M + S	33	34
M + S <= X-BAR < M + 2S	11	13 or 14
M + 2S <= X-BAR < M + 3S	3	2 or 3

Click on the graph button.

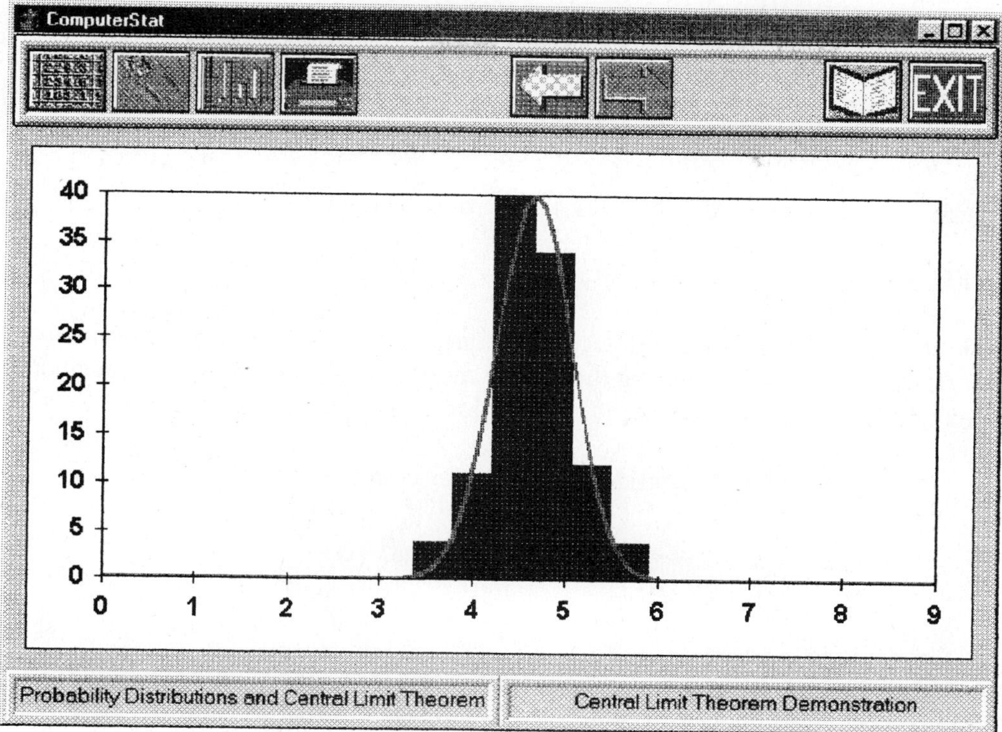

Lab Activities for Central Limit Theorem

These activities coordinate with Section 7.2 of *Understandable Statistics*

Using the *ComputerStat* program CENTRAL LIMIT THEOREM, compare the distribution of sample means taken from 100 samples of size N with the results predicted by the Central Limit Theorem. Look at the results when N is 2, 5, 10, 20, 30, 40, 50. In each case compare the mean of the $\bar{x}$ distribution with the predicted mean. Compare the tallies of values falling in the specified intervals with those predicted by the Central Limit Theorem. Finally, look at the graphs. Notice what happens as the sample size becomes larger and larger. In general, can we say that the larger the sample size, the narrower the $\bar{x}$ distribution?

CHAPTER 8 ESTIMATION

CONFIDENCE INTERVAL DEMONSTRATION

(Section 8.1 of *Understandable Statistics*)

Main menu selection: CONFIDENCE INTERVALS

Sub-menu selection: CONFIDENCE INTERVAL DEMONSTRATION

I. Description of the Program

This program constructs 90% confidence intervals for the population mean μ of random numbers from 0 to 1. Each interval is found using the sample mean $\bar{x}$ and sample standard deviation s computed from a sample of N random numbers from 0 to 1.

A 90% confidence interval for μ is

$$\bar{x} - 1.645\frac{s}{\sqrt{n}} \quad \text{to} \quad \bar{x} + 1.645\frac{s}{\sqrt{n}}$$

The computer will display 20 intervals, graphs of the intervals, and a cumulative tally and corresponding percent of the number of confidence intervals that actually contain μ = 0.5. This percent should approach 90 as the total number of confidence intervals we examine of a given sample size becomes larger and larger.

Input: Sample size N; N between 30 and 500 inclusive

Output: Table showing sample means, standard deviations and 90% confidence intervals for 20 different random samples. The table also indicates whether or not the confidence interval contains the population mean μ = 0.5

A graphical display of the 20 confidence intervals described in the table.

Summary indicating the number of confidence intervals of the same sample size and percent which contain the population mean μ.

II. Sample Run

Generate twenty 90% confidence intervals from samples of size N = 35. Recall that our population consists of the numbers between 0 and 1, so the population mean = 0.5. Enter the value of 35 for N and click on the calculate button.

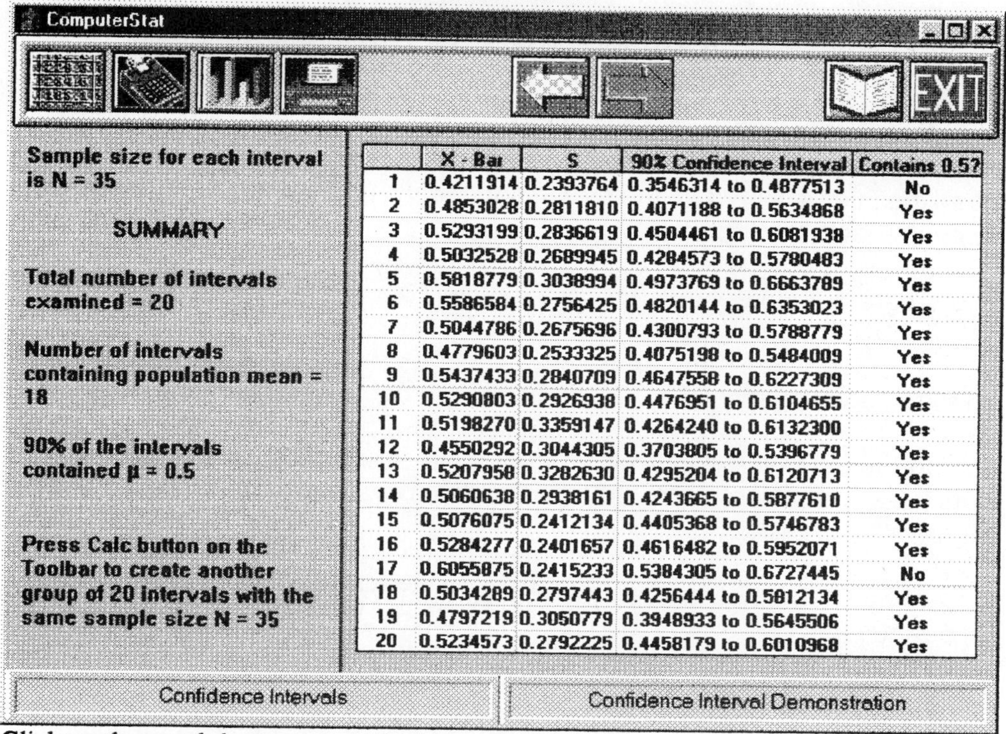

ComputerStat

EXIT

Sample size for each interval is N = 35

SUMMARY

Total number of intervals examined = 20

Number of intervals containing population mean = 18

90% of the intervals contained μ = 0.5

Press Calc button on the Toolbar to create another group of 20 intervals with the same sample size N = 35

	X - Bar	S	90% Confidence Interval	Contains 0.5?
1	0.4211914	0.2393764	0.3546314 to 0.4877513	No
2	0.4853028	0.2811810	0.4071188 to 0.5634868	Yes
3	0.5293199	0.2836619	0.4504461 to 0.6081938	Yes
4	0.5032528	0.2689945	0.4284573 to 0.5780483	Yes
5	0.5818779	0.3038994	0.4973769 to 0.6663789	Yes
6	0.5586584	0.2756425	0.4820144 to 0.6353023	Yes
7	0.5044786	0.2675696	0.4300793 to 0.5788779	Yes
8	0.4779603	0.2533325	0.4075198 to 0.5484009	Yes
9	0.5437433	0.2840709	0.4647558 to 0.6227309	Yes
10	0.5290803	0.2926938	0.4476951 to 0.6104655	Yes
11	0.5198270	0.3359147	0.4264240 to 0.6132300	Yes
12	0.4550292	0.3044305	0.3703805 to 0.5396779	Yes
13	0.5207958	0.3282630	0.4295204 to 0.6120713	Yes
14	0.5060638	0.2938161	0.4243665 to 0.5877610	Yes
15	0.5076075	0.2412134	0.4405368 to 0.5746783	Yes
16	0.5284277	0.2401657	0.4616482 to 0.5952071	Yes
17	0.6055875	0.2415233	0.5384305 to 0.6727445	No
18	0.5034289	0.2797443	0.4256444 to 0.5812134	Yes
19	0.4797219	0.3050779	0.3948933 to 0.5645506	Yes
20	0.5234573	0.2792225	0.4458179 to 0.6010968	Yes

Confidence Intervals

Confidence Interval Demonstration

Click on the graph button to see the intervals displayed.

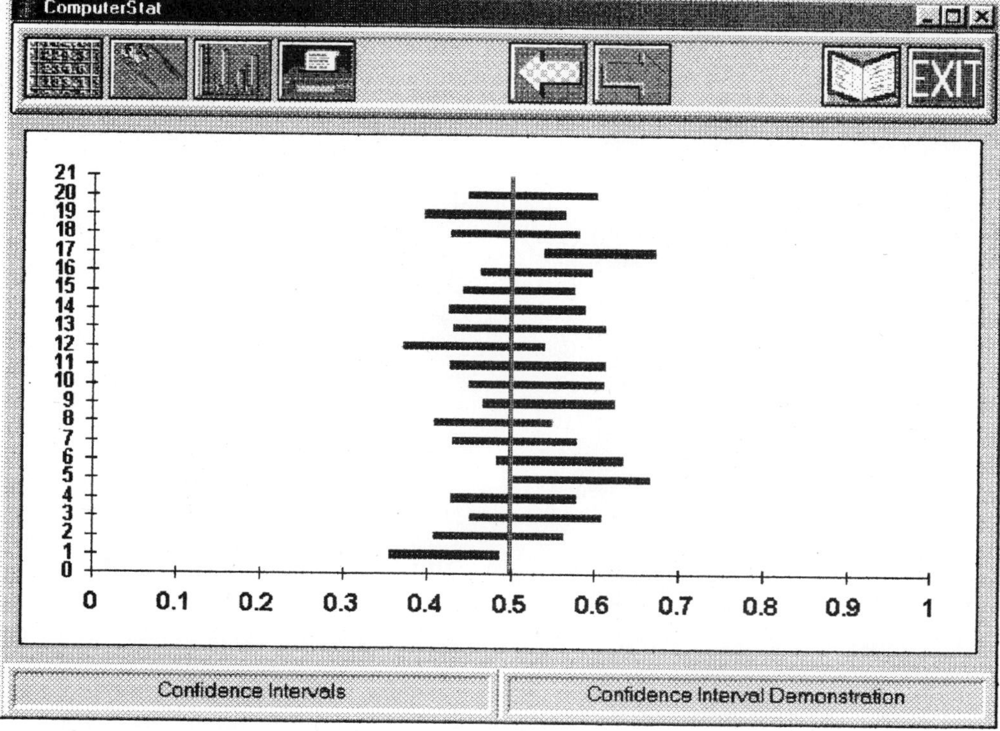

Lab Activities Using Confidence Interval Demonstration

These activities coordinate with Section 8.1 of *Understandable Statistics.*

1. Run the program with sample of size 35 a total of 5 times to obtain 100 confidence intervals.

   a) Are each of the intervals the same? Why or why not? Note that the sample means and standard deviations of all samples are slightly different.

   b) Do all the intervals contain the population mean of 0.5? In the long run, about what percent should contain 0.5?

2. Run the program for intervals from sample size 30 and then for intervals with sample size 50.

   a) Do the intervals seem to be shorter when we use a sample of size 50?

   b) Do you predict that the intervals for samples of size 100 will be shorter or longer than those for sample size 50? Run the program with N = 100 to verify your answer.

   c) In general, how can you narrow the confidence interval without changing the confidence level?

CONFIDENCE INTERVALS FOR A POPULATION MEAN MU
(Sections 8.1 and 8.2 of *Understandable Statistics*)

Main menu selection: CONFIDENCE INTERVALS

Sub-menu selection CONFIDENCE INTERVALS FOR A POPULATION MEAN MU

I. Description of the Program

This program finds confidence intervals for a population mean μ. The user may
a) Select one of six Class Demonstrations
b) Enter sample raw data
c) Enter processed data; the sample mean $\overline{x}$ and sample standard deviation s

Small samples (sample size less than 30) use critical values from the inverse Student's t distribution. Large samples use critical values from the inverse normal distribution. The user inputs a confidence level C and the computer returns endpoints of the C% confidence interval.

Input: Data entry

Option 1: Classroom Demonstrations

ComputerStat

EXIT

CLASS DEMONSTRATION #1: Number of Wolf Pups in a Den

CLASS DEMONSTRATION #2: Weights of Pro Football Players

CLASS DEMONSTRATION #3: Heights of Pro Basketball Players

CLASS DEMONSTRATION #4: Miles per Gallon Gasoline Consumption

CLASS DEMONSTRATION #5: Fasting Glucose Blood Tests for a Specific Patient

CLASS DEMONSTRATION #6: Number of Children in Rural Canadian Families

Enter Your Own Data

Confidence Intervals | Confidence Intervals for a Population Mean μ

Option 2:

a) Enter raw data with data

b) Enter processed data: That is specify sample size N, mean of the data $\overline{x}$ and sample standard deviation s

Sample size, N; for raw data N must be between 2 and 100 inclusive

Confidence level, C: C as a percent greater than 1 and less than 100

Output: Computer tells you if you have a large or small sample

Information summary: confidence level, sample size, sample mean, sample standard deviation

Critical value of Z_0 for large samples

Critical value of t_0 for small samples

The C% confidence interval for the population mean μ

II. Sample Run

Start the program and select Classroom Demonstration #1: NUMBER OF WOLF PUPS IN A DEN. Use a 90% confidence level.

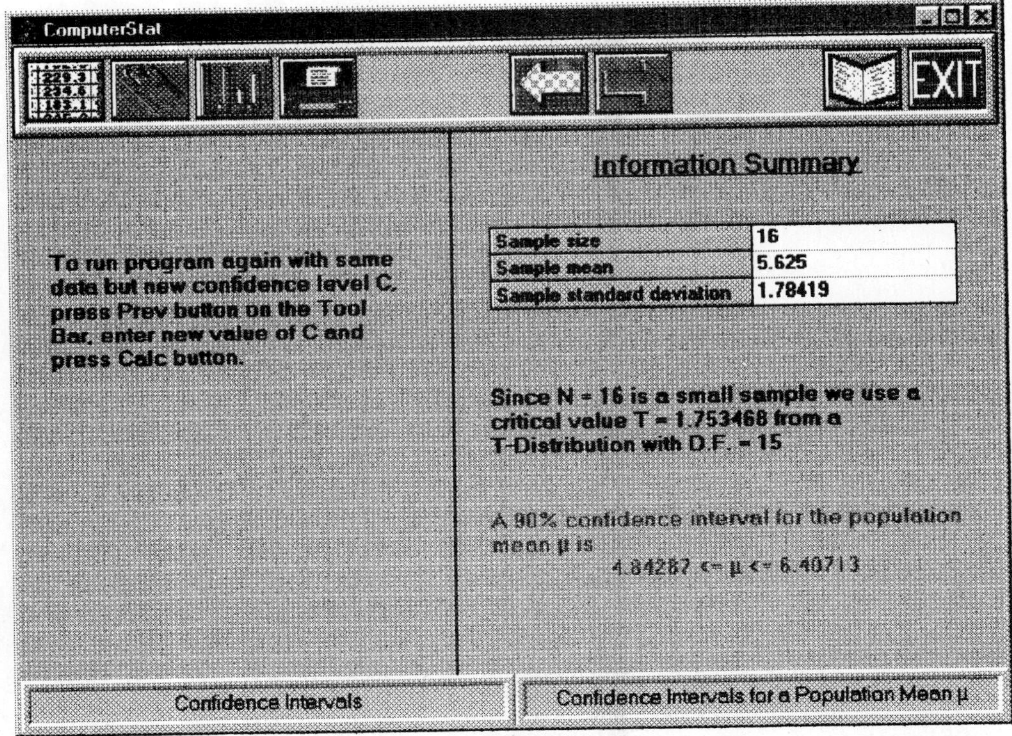
ComputerStat
EXIT
Information Summary
Sample size
16
Sample mean
5.625
Sample standard deviation
1.78419
To run program again with same data but new confidence level C, press Prev button on the Tool Bar, enter new value of C and press Calc button.
Since N = 16 is a small sample we use a critical value T = 1.753468 from a T-Distribution with D.F. = 15
A 90% confidence interval for the population mean μ is
4.84287 <= μ <= 6.40713
Confidence Intervals
Confidence Intervals for a Population Mean μ

Lab Activities for CONFIDENCE INTERVALS FOR A POPULATION MEAN MU

These activities coordinate with Sections 8.1 and 8.2 of *Understandable Statistics.*

1. Market Survey was hired to do a study for a new soft drink, Refresh. A random sample of 20 people were given a can of Refresh and asked to rate it for taste on a scale of 1 to 10 (with 10 being the highest rating). The ratings were

5	8	3	7	5	9	10	6	6	2
9	2	1	8	10	2	5	1	4	7

Find an 85% confidence interval for the population mean rating of Refresh.

2. Select CLASS DEMONSTRATION #3: HEIGHTS OF PRO BASKETBALL PLAYERS.

   a) Find a 99% confidence interval for the population mean height.
   b) Find a 95% confidence interval for the population mean height.
   c) Find a 90% confidence interval for the population mean height.
   d) Find an 85% confidence interval for the population mean height.
   e) What do you notice about the length of the confidence interval as the confidence level goes down? If you used a confidence level of 80%, do you expect the confidence interval to be longer or shorter than that of 85%? Run the program again to verify your answer.

3. Under the option to enter your own data, you can choose to enter only the sample size N, the sample mean $\bar{x}$ and the sample standard deviation s for a data set. Select this option. Suppose your read newspaper article that reported a random sample of U.S. income tax returns showed that the average time for a refund to be issued is 45 days with sample standard deviation of 8 days.

   a) If the sample size N = 20, find a 95% confidence interval for the population mean number of days required to issue a refund.
   b) If the sample size N = 40, find a 95% confidence interval for the population mean number of days required to issue a refund.
   c) If the sample size N = 60, find a 95% confidence interval for the population mean number of days required to issue a refund.
   d) If the sample size N = 100, find a 95% confidence interval for the population mean number of days required to issue a refund.
   e) What do you notice about the length of the confidence interval as the sample size increases? If you used a sample size of N = 500, do you expect the confidence interval to be longer or shorter than when you use a sample size of 100? Run the program again with N = 500 to verify your answer.

CONFIDENCE INTERVALS FOR THE PROBABILITY OF SUCCESS P IN A BINOMIAL DISTRIBUTION

(Section 8.3 of *Understandable Statistics*)

Main menu selection: CONFIDENCE INTERVALS

Sub-menu selection: CONFIDENCE INTERVALS FOR THE PROBABILITY OF SUCCESS IN A BINOMIAL DISTRIBUTION

I. Description of the program

This program finds confidence intervals for p, the probability of success on a single trial of a binomial distribution. We assume that the number of trials N is large enough to permit the use of a normal approximation to the binomial distribution. Empirical studies show that the normal approximation is reasonable if both NP and N(1-P) are both greater than 5. The computer checks that these conditions are met

Input: The percent confidence level, C; C as a percent between 1 and 100

Number of trials, N; N greater than or equal 12.

Number of successes out of N trials, R; R between 0 and N inclusive

Output: Information summary listing percent confidence level C, number of trials, sample approximation $\hat{p}$ for P, comment indicating if a normal approximation to the binomial is appropriate

The confidence interval for P

II. Sample Run

The public television station BPBS wants to find the percent of its viewing population who give donations to the station. A random sample of 300 viewers were surveyed and it was found that 123 made contributions to the station. Find a 95% confidence interval for the probability that a viewer of BPBS selected at random contributes to the station.

Select the program. The program begins with its title and brief description. Since we have binomial trials, this program is appropriate.

Enter the value 95 for the percent confidence level. Use the value N = 300 since there are 300 viewers in the sample. The number of successes is R = 123. The output follows.

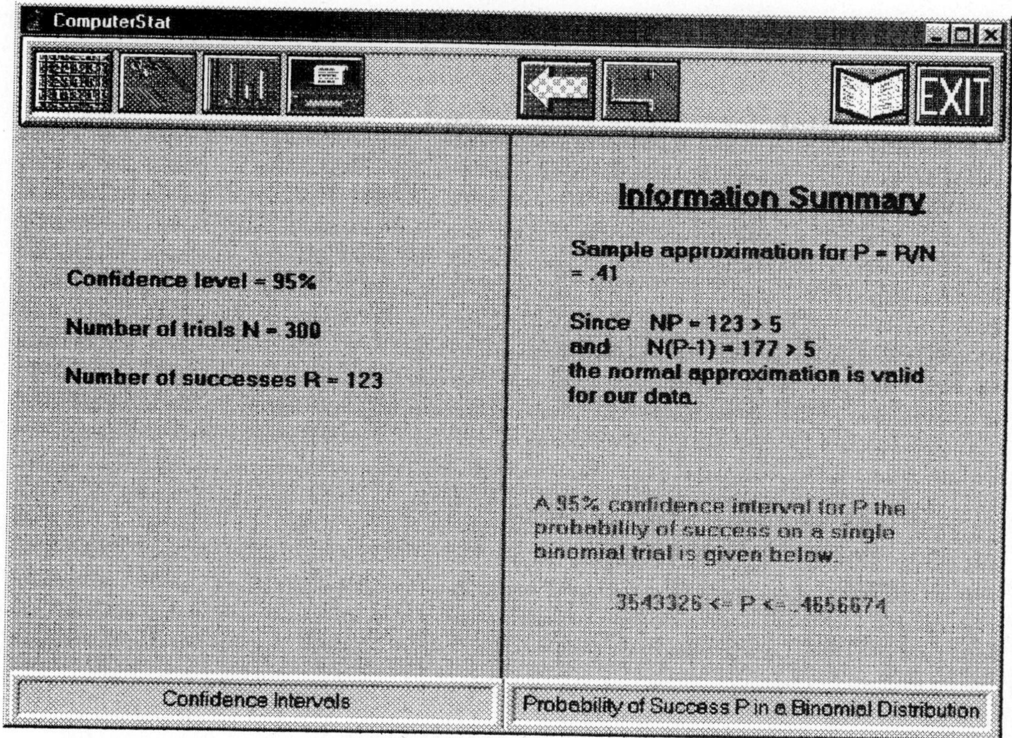

Lab Activities for CONFIDENCE INTERVALS FOR THE PROBABILITY OF SUCCESS P IN A BINOMIAL DISTRIBUTION

These activities coordinate with Section 8.3 of *Understandable Statistics*.

1. There are many types of errors that will cause a computer program to terminate or give incorrect results. One type of error is punctuation. For instance, if a comma is inserted in the wrong place, the program might not run. A study of programs written by students in a beginning programming course showed that 75 out of 300 errors selected at random were punctuation errors. Find a 99% confidence interval for the proportion of errors made by beginning programming students that are punctuation errors. Next find a 90% confidence interval. Is this interval longer or shorter?

2. Sam decided to do a statistics project to determine a 90% confidence interval for the probability that a student at West Plains College eats lunch in the school cafeteria. He surveyed a random sample of 12 students and found that 9 ate lunch in the cafeteria. Can Sam use the program to find a confidence interval for the population proportion of students eating in the cafeteria? Why or why not? Try the program with N = 12 and R = 9. What happens? What should Sam do to complete his project?

CONFIDENCE INTERVALS FOR MU1-MU2 (INDEPENDENT SAMPLES)
CONFIDENCE INTERVALS FOR P1-P2 (LARGE SAMPLES)
(Section 8.5 of *Understandable Statistics*)

Main menu selection: CONFIDENCE INTERVALS

Sub-menu selection: CONFIDENCE INTERVALS FOR MU1-MU2
CONFIDENCE INTERVALS FOR P1-P2

I. Description of the Programs

These programs find C% confidence intervals for the difference of means (independent samples and difference of proportions respectively. Confidence intervals for difference of means are found for both large and small independent samples. Confidence intervals for difference of proportions assume that the number of trials for each proportion are large enough to justify the use of the normal approximation to the binomial distribution.

Input:

difference of means

Data entry:

Option 1: Select a class demonstration

Option 2: Enter your own data
- a) Enter the raw data for each distribution
- b) Enter processed data: Sample size, sample man, and sample standard deviation for each distribution

Percent confidence level C: C greater than 1 and less than 100

difference of proportions

Number of trials for first distribution, N1; N1 at least 12
Number of successes for first distribution, R1
Number of trials for second distribution, N2; N2 at least 12
Percent confidence level C: C greater than 1 and less than 100

Output:

Information summary
The confidence interval for the difference

II. Sample Run (Difference of Means)

Select Class Demonstration #1 to find a 95% confidence interval for the difference of means in heights of Pro football players and Pro basketball players. Does the confidence interval indicate that the mean height of basketball players is greater than the mean height of football players?

Start the program and select Class Demonstration #1.

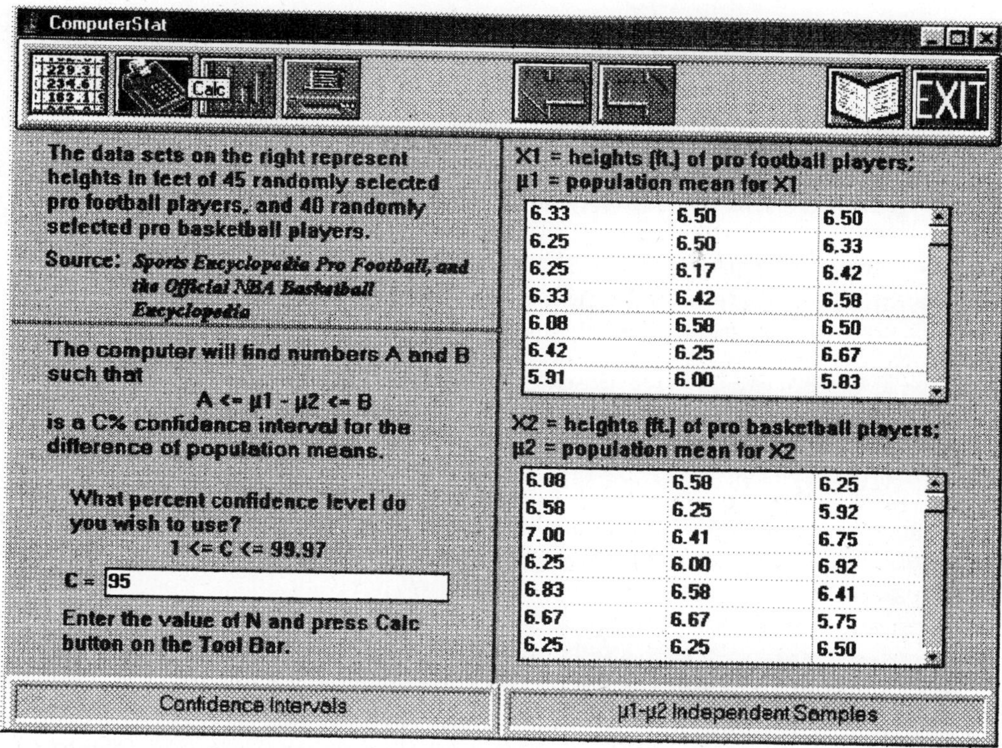
ComputerStat
Calc
EXIT
The data sets on the right represent heights in feet of 45 randomly selected pro football players, and 40 randomly selected pro basketball players.
Source: Sports Encyclopedia Pro Football, and the Official NBA Basketball Encyclopedia
The computer will find numbers A and B such that
A <= µ1 - µ2 <= B
is a C% confidence interval for the difference of population means.
What percent confidence level do you wish to use?
1 <= C <= 99.97
C = 95
Enter the value of N and press Calc button on the Tool Bar.
X1 = heights (ft.) of pro football players;
µ1 = population mean for X1
6.33 6.50 6.50
6.25 6.50 6.33
6.25 6.17 6.42
6.33 6.42 6.58
6.08 6.58 6.50
6.42 6.25 6.67
5.91 6.00 5.83
X2 = heights (ft.) of pro basketball players;
µ2 = population mean for X2
6.08 6.58 6.25
6.58 6.25 5.92
7.00 6.41 6.75
6.25 6.00 6.92
6.83 6.58 6.41
6.67 6.67 5.75
6.25 6.25 6.50
Confidence Intervals
µ1-µ2 Independent Samples

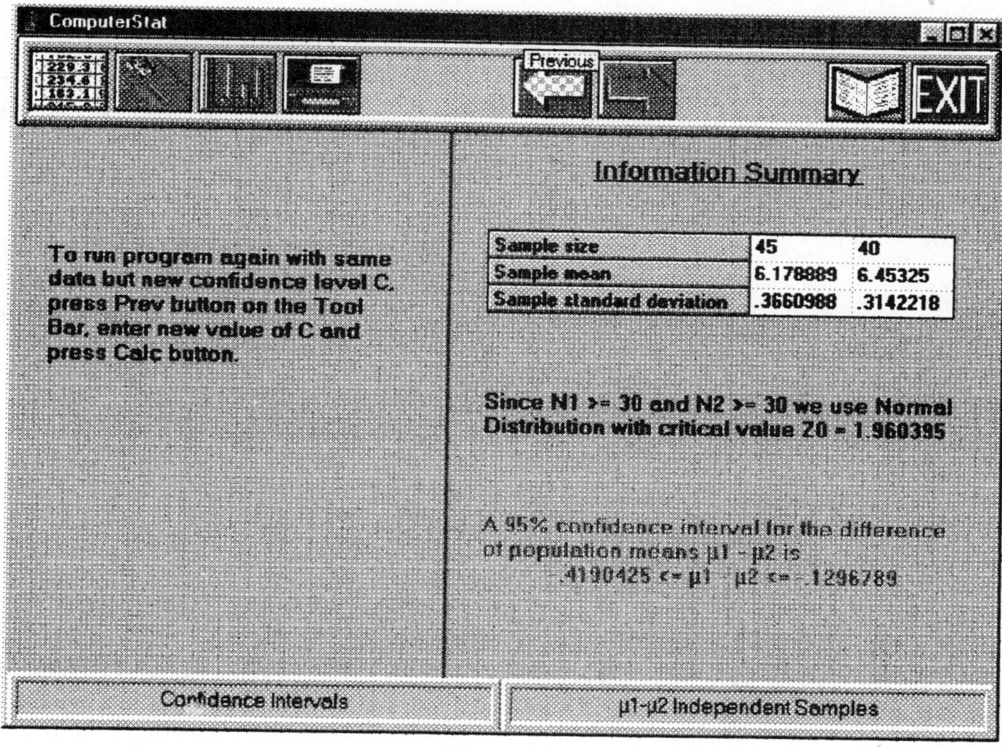
ComputerStat
Previous
EXIT
To run program again with same data but new confidence level C, press Prev button on the Tool Bar, enter new value of C and press Calc button.
Information Summary
Sample size 45 40
Sample mean 6.178889 6.45325
Sample standard deviation .3660988 .3142218
Since N1 >= 30 and N2 >= 30 we use Normal Distribution with critical value Z0 = 1.960395
A 95% confidence interval for the difference of population means µ1 - µ2 is
-.4190425 <= µ1 - µ2 <= -.1296789
Confidence Intervals
µ1-µ2 Independent Samples

Since the confidence interval contains numbers that are all negative, it indicates that the mean heights of the basketball players μ_2 is greater than the mean height of football players μ_1.

Lab Activities for Confidence Intervals for MU1- MU2 or for P1-P2

These activities coordinate with Section 8.5 of *Understandable Statistics.*

1. Run CONFIDENCE INTERVALS FOR MU1-MU2. Select Class Demonstration #4 Number of Cases of Red Fox Rabies in Two Regions of Germany.

    a) Use a confidence level C% = 90. Does the interval indicate that the population mean number of rabies in region 1 is greater than the population mean number of rabies in region 2 at the 90% level? Why or why not?
    b) Use a confidence level C% = 95. Does the interval indicate that the population mean number of rabies in region 1 is greater than the population mean number of rabies in region 2 at the 95% level? Why or why not?
    c) Predict whether or not the population mean number of rabies in region 1 is greater than the population mean number of rabies in region 2 at the 99% level. Explain your answer. Verify your answer by running the program with a 99% confidence level

2. A random sample of 30 police officers working the night shift showed that 23 used at least 5 sick leave days per year. Another random sample of 45 police officers working the day shift showed that 26 used at least 5 sick leave days per year. Use the program CONFIDENCE INTERVAL FOR P1-P2 to find a 90%confidence interval for the difference of population proportions of police officers working the two shifts using at least 5 sick leave days per year. At the 90% level, does it seems that the proportion of officers working the night shift who use at least 5 sick leave days per year is higher than the proportion of day shift officers using that much leave? Does there seem to be a difference in proportions at the 99% level? Why or why not?

CHAPTER 9 HYPOTHESIS TESTING

TESTING A SINGLE POPULATION MEAN
(Sections 9.1, 9.2, 9.3, 9.4 of *Understandable Statistics*)

Main menu selection: HYPOTHESIS TESTING

Sub-menu selection: TESTING A SINGLE POPULATION MEAN

I. Description of the Program

Let MU be the population mean of a distribution of x values. We will assume the x distribution is one for which both the population mean and standard deviation are unknown. In the case in which we have only a small sample we also must assume that the x distribution is approximately a normal distribution. This program tests the null hypothesis

H0:MU = K

against the alternate hypothesis which may be

H1:MU < K (left tail test)
H1:MU > K (right tail test)
H1:MU <> K (two tail test)

The program concludes the hypothesis test using P values as well as the critical region and sample test statistic.

Input: Data entry

Option 1: Use a Class Demonstration

<<<<< OPTIONS AVAILABLE >>>>>

1) CLASS DEMONSTRATION #1: NUMBER OF WOLF PUPS IN A DEN
2) CLASS DEMONSTRATION #2: WEIGHTS OF PRO FOOTBALL PLAYERS
3) CLASS DEMONSTRATION #3: HEIGHTS OF PRO BASKETBALL PLAYERS
4) CLASS DEMONSTRATION #4: MILES PER GALLON GASOLINE CONSUMPTION
5) CLASS DEMONSTRATION #5: FASTING GLUCOSE BLOOD TESTS FOR A SPECIFIC PATIENT
6) CLASS DEMONSTRATION #6: NUMBER OF CHILDREN IN RURAL CANADIAN FAMILIES

Option 2: Enter data
- a) Enter raw data with data
- b) Enter processed data, that is: sample size, sample mean, sample standard deviation

The constant K in the null hypothesis H0:MU = K

Type of test: H1:MU < K (left tail)
H1:MU > K (right tail)
H1:MU <> K (two tail)

Level of significance, alpha; alpha between 0 and 0.5

Output: Information summary table listing sample size, sample mean, sample standard deviation, level of significance, null hypothesis, alternate hypothesis, P value Test conclusion with mention of distribution used, T or Z value corresponding to the sample test statistic $\bar{x}$, location of sample test statistic inside or outside the critical region and P value compared to the level of significance alpha.

II. Sample Run

Start the program and select Class Demonstration #3 HEIGHTS OF PRO BASKETBALL PLAYERS. Notice that the level of significance and hypotheses are already given. These may be changed later. At the alpha = 0.05 level of significance, do the data support the conclusion that the average height of the players is different from 6.5 feet?

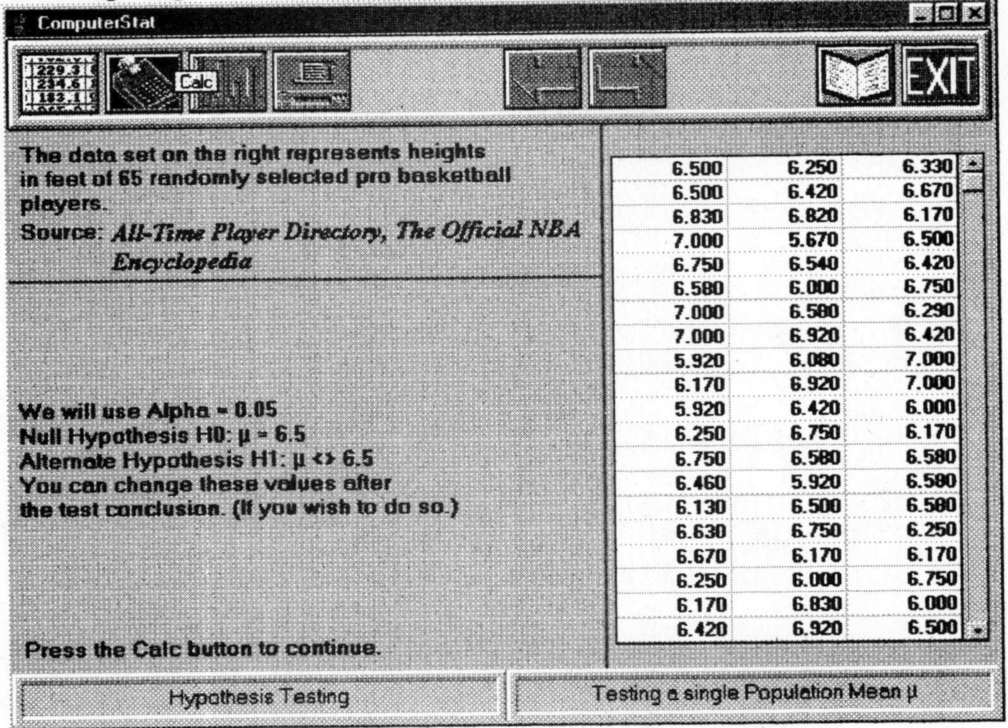

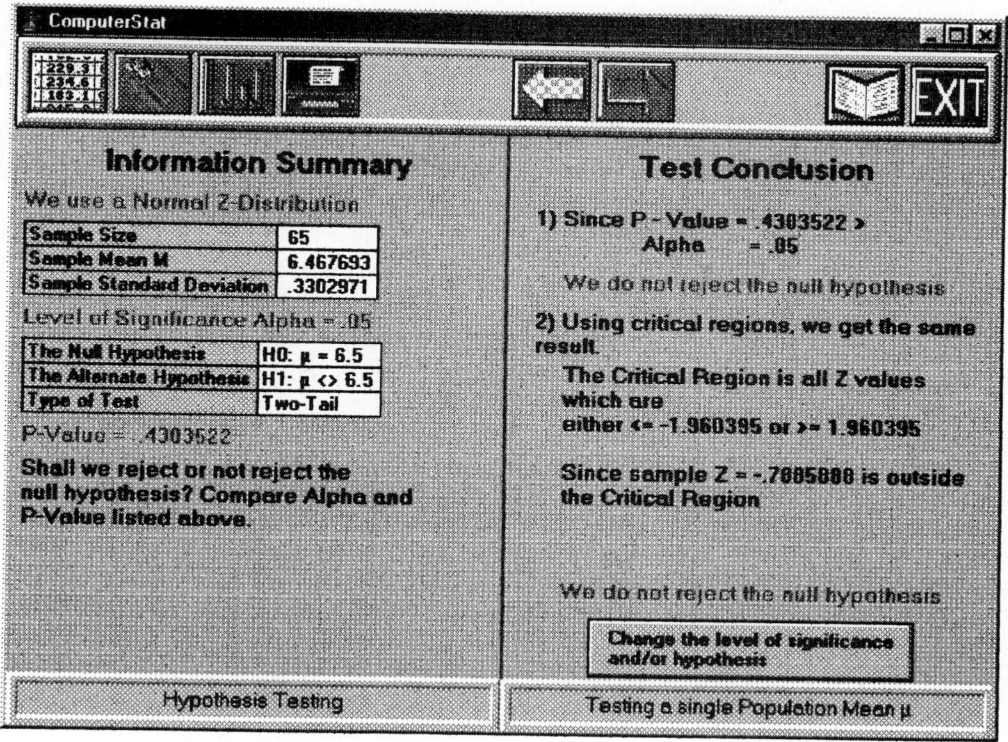

Lab Activities Using TESTING A SINGLE POPULATION MEAN

These activities coordinate with Section 9.2, 9.3 and 9.4 of *Understandable Statistics*

1. Select Class Demonstration #3 regarding the heights of pro basketball players. In the conlcusion window, select the option to change the hypothesis and/or level of significance.

   This time test the hypothesis that the average height of the players is greater than 6.2 feet at the 1% level of significance. To do this, change the value of K to 6.2 and the alternate hypothesis to a right tail test. Change the level of significance to 0.01.

2. In this problem we will see how the test conclusion is possibly affected by a change in the level of significance.

   Teachers for Excellence is interested in the attention span of students in grades 1 and 2 now as compared to 20 years ago. They believe it has decreased. Studies done 20 years ago indicate that the attention span of children in grades 1 and 2 was 15 minutes. A study sponsored by Teachers for Excellence involved a random sample of 20 students in grades 1 and 2. The average attention span of these students was (in minutes) $\overline{x} = 14.2$ with standard deviation $s = 1.5$.

   a) Use the program to conduct the hypothesis test using $\alpha = 0.05$ and a left tailed test. What is the test conclusion? What is the P value?

Lab Activities Using TESTING A SINGLE POPULATION MEAN continued

b) Use the program to conduct the hypothesis test using $\alpha = 0.01$ and a left tailed test.. What is the test conclusion? How could you have predicted this result by looking at the P value from part (a)? Is the P value for this part the same as it was for part (a)?

3. In this problem, let's explore the effect that sample size has on the process of testing a mean. Run the program with the hypotheses H_0: $\mu = 200$, H_1: $\mu > 200$, $\alpha = 0.05$, $\bar{x} = 210$ and $s = 40$.

   a) Use the sample size $N = 30$. Note the P-value, and Z score of the sample test statistic and test conclusion.
   b) Use the sample size $N = 50$. Note the P-value, and Z score of the sample test statistic and test conclusion.
   c) Use the sample size $N = 100$. Note the P-value, and Z score of the sample test statistic and test conclusion.
   d) In general, if your sample statistic is close to the proposed population mean specified in H_0, and you want to reject H_0, would you use a smaller or a larger sample size?

OTHER TESTS OF HYPOTHESIS
(Sections 9.5, 9.6, 9.7 of *Understandable Statistics*)

ComputerStat contains programs for all the other hypothesis tests in Chapter 9 of *Understandable Statistics*.

Main menu selection: HYPOTHESIS TESTING

Sub-menu selections: **TESTING A SINGLE PROPORTION P**
(Section 9.5 of *Understandable Statistics*)
TESTING MU1-MU2 (DEPENDENT SAMPLES)
(Section 9.6 of *Understandable Statistics*)
TESTING MU1-MU2 (INDEPENDENT SAMPLES)
(Section 9.7 of *Understandable Statistics*)
TESTING P1-P2 (LARGE SAMPLES)
(Section 9.7 of *Understandable Statistics*)

I. Description of the Programs

All of these programs are similar to the program TESTING A SINGLE MEAN MU in the sense that you enter the target value K for the null hypothesis, select the alternate hypothesis, and specify the level of significance alpha. Each program displays an information summary, the Z or T value corresponding to the sample test statistic, the P value corresponding to the sample test statistic, and the test conclusion.

Input: All the programs have options to enter your own data.

Testing MU1-MU2 (Dependent Samples) has these Class Demonstrations

1) CLASS DEMONSTRATION #1: ANNUAL SALARIES FOR MALE VERSUS FEMALE ASSISTANT PROFESSORS
2) CLASS DEMONSTRATION #2: UNEMPLOYMENT PERCENTAGE FOR HIGH SCHOOL ONLY VERSUS COLLEGE GRADUATES
3) CLASS DEMONSTRATION #3: NUMBER OF TRADITIONAL NAVAJO HOGANS VERSUS NUMBER OF HOUSES ON THE NAVAJO INDIAN RESERVATION
4) CLASS DEMONSTRATION #4: AVERAGE MONTHLY TEMPERATURE IN MIAMI VERSUS HONOLULU

Testing MU1-MU2 (Independent Samples) has these Class Demonstrations

1) CLASS DEMONSTRATION #1: HEIGHTS OF PRO FOOTBALL PLAYERS VERSUS HEIGHTS OF PRO BASKETBALL PLAYERS
2) CLASS DEMONSTRATION #2: WEIGHTS OF PRO FOOTBALL PLAYERS VERSUS WEIGHTS OF PRO BASKETBALL PLAYERS
3) CLASS DEMONSTRATION #3: SEPAL WIDTH OF IRIS VERSICOLOR VERSUS IRIS VIRGINICA

4) CLASS DEMONSTRATION #4: NUMBER OF CASES OF RED FOX RABIES IN TWO REGIONS OF GERMANY

Output: Information summary
Z or T value of the sample test statistic
P value of the sample test statistic
Test conclusion

Lab Activities Using Other Tests of Hypothesis

These activities coordinate to Sections 9.5, 9.6, and 9.7 of *Understandable Statistics.*

1. Select TESTING MU1-MU2 (DEPENDENT SAMPLES) and use one of the class demonstrations.
   a) Use alpha = 0.01 with a two tailed test and note the P-value and test conclusion.
   b) Use alpha = 0.05 with a two tailed test and note the P-value and test conclusion.
   c) Does the P value change? Does the sample test statistic change? Does the critical region change? Does the test conclusion change?

2. Select TESTING MU1-MU2 (INDEPENDENT SAMPLES) and use one of the class demonstrations.
   a) Use alpha = 0.01 with a two tailed test and note the value of the sample test statistic $\bar{x}_1 - \bar{x}_2$ and test conclusion.
   b) Select the same class demonstration from CONFIDENCE INTERVALS FOR MU1-MU2 (INDEPENDENT SAMPLES). Find a 99% confidence interval and record the endpoints of the interval.
   c) Is the value of the sample test statistic $\bar{x}_1 - \bar{x}_2$ found in part (a) inside or outside the confidence interval? If it is inside, was the test conclusion of part (a) "Do not Reject H_0"? If it is outside was the test conclusion "Reject H_0"?

3. Jones Computer Security is testing a new security device which is believed to decrease the incidence of computer "break ins." Without this device, the computer security test team can break security 47% of the time. With the device in place, the test team made 400 attempts and were successful 82 times. Select an appropriate program from the HYPOTHESIS TESTING menu and test the claim that the device reduces the proportion of successful break ins. Use alpha = 0.05 and note the P-value. Does the test conclusion change for alpha = 0.01?

Lab Activities Using Other Tests of Hypothesis continued

4. Publisher's Survey did a study to see if the proportion of men who read mysteries is different than the proportion of women who read them. A random sample of 402 women showed that 112 read mysteries regularly (at least 6 books per year). A random sample of 365 men showed that 92 read mysteries regularly. Is the proportion of mystery readers different between men and women? Use a 1% level of significance.

   a) Look at the P value of the test conclusion. Jot it down.
   b) Test the hypothesis that the proportion of women who read mysteries is *greater* than the proportion of men. Use a 1% level of significance. Is the P value for a right tailed test half that of a two tailed test? If you know the P-value for a two tailed test, can you draw conclusions for a one tailed test?

5. The programs of *ComputerStat* offer a tremendous aid in performing the computations required for hypothesis testing. However, the user must identify which program to use, select the hypotheses, and interpret the test conclusions in terms of the original statement of a problem. Select appropriate programs from *ComputerStat* to do the even numbered problems of Chapter 9 Review Problems in *Understandable Statistics.*

CHAPTER 10 REGRESSION AND CORRELATION

LINEAR REGRESSION AND CORRELATION
(Sections 10.1, 10.2, 10.3, and 10.4 of *Understandable Statistics*)

Main menu selection:	LINEAR REGRESSION AND CORRELATION
Sub-menu selection:	LINEAR REGRESSION AND TESTING THE CORRELATION COEFFICIENT RHO

Main menu selection:	HYPOTHESIS TESTING
Sub -menu selection:	LINEAR REGRESSION AND TESTING THE CORRELATION COEFFICIENT RHO

I. Description of the Program

The program LINEAR REGRESSION AND CORRELATION (performs a number of functions related to linear regression and correlation. The user inputs a random sample of ordered pairs of X and Y values. Then the computer returns the sample means, sample standard deviations of X and Y values; largest and smallest X value; standard error of estimate; equation of least squares line; the correlation coefficient and the coefficient of determination.

Next the user has several options.

1) Predict Y values from X values, and when the X value is in the proper range, compute a C% confidence intervals for Y values.

2) Test the correlation coefficient. If we let RHO represent the population correlation coefficient, then the null hypothesis is H0:RHO = 0. The alternate hypothesis may be one of the following

   H1:RHO < 0 (left tailed test)
   H1:RHO > 0 (right tailed test)
   H1:RHO <>0 (two tailed test)

3) Graph data points (X,Y) to create a scatter diagram. Then the user can see five points on the least squares line.

Input: Option 1: Select a class demonstration from the data button menu itens
1) CLASS DEMONSTRATION #1: LIST PRICE VERSUS BEST PRICE FOR A NEW GMC PICKUP TRUCK
2) CLASS DEMONSTRATION #2: CRICKET CHIRPS VERSUS TEMPERATURE
3) CLASS DEMONSTRATION #3: DIAMETER OF SAND GRANULES VERSUS SLOPE ON A NATURALLY OCCURRING OCEAN BEACH
4) CLASS DEMONSTRATION #4: NATIONAL UNEMPLOYMENT RATE MALE VERSUS FEMALE

Option 2: Enter your own data with data correction possible

Value of X for which you want to predict Y
Confidence level C for confidence interval of Y values
Alternate hypothesis for testing RHO
Level of significance for testing RHO

Output: Information summary table including the least squares line, S_e, r, r^2
For a given X value, the predicted Y value and a C% confidence interval
Test summary for testing RHO
Scatter diagram and least squares line

II. Sample run

Begin the program Linear Regression and Testing the Correlation Coefficient Rho. Click on the data button. Select Class Demonstration #2) CRICKET CHIRPS VERSUS TEMPERATURE. Use the program to explore the relationship between cricket chirp frequency and temperature.

After you select the class demonstration, the data is shown followed by the information summary

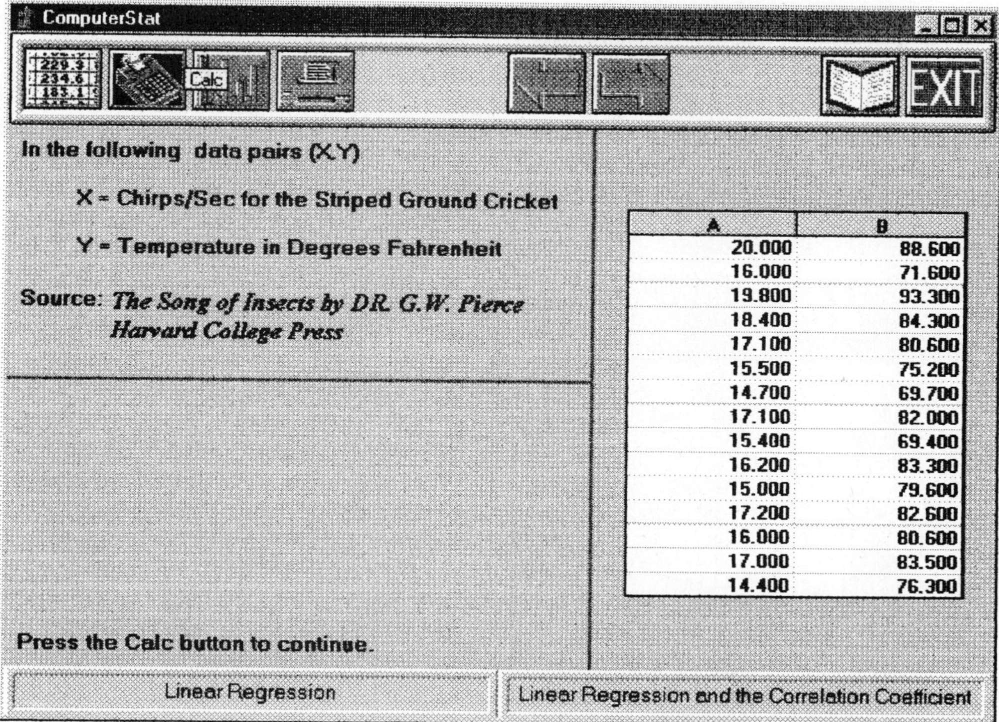

Click on the calculate button.

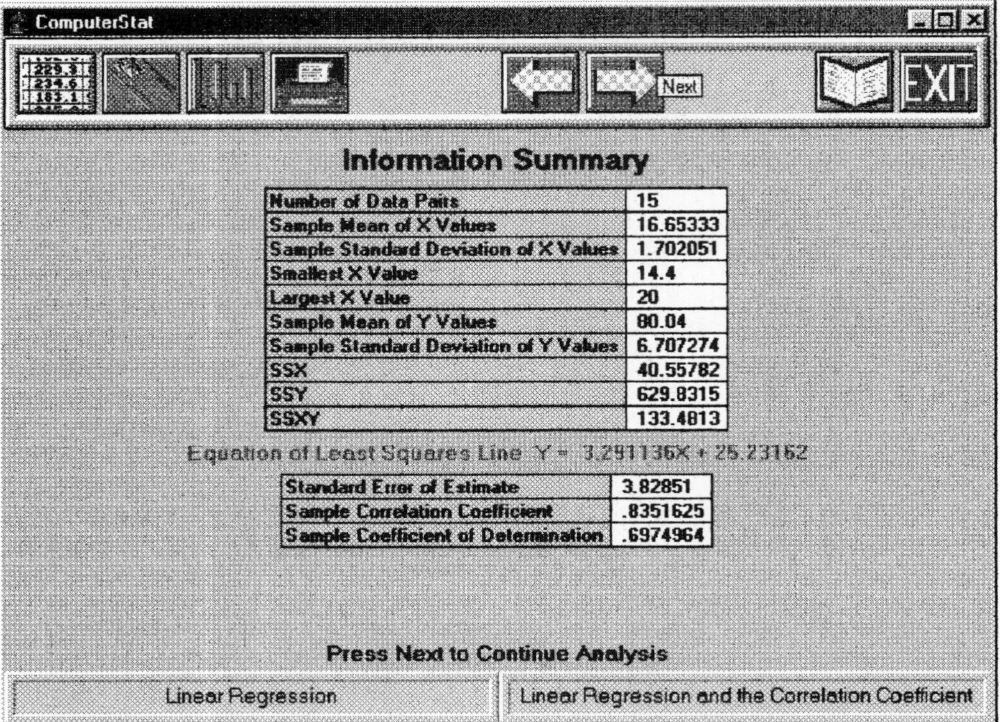

Clicking on the Next button gives you a choice of graphing the scatter plot and least squares line, or predicting y from an x value or testing the correlation coefficient r. Click on the graph button.

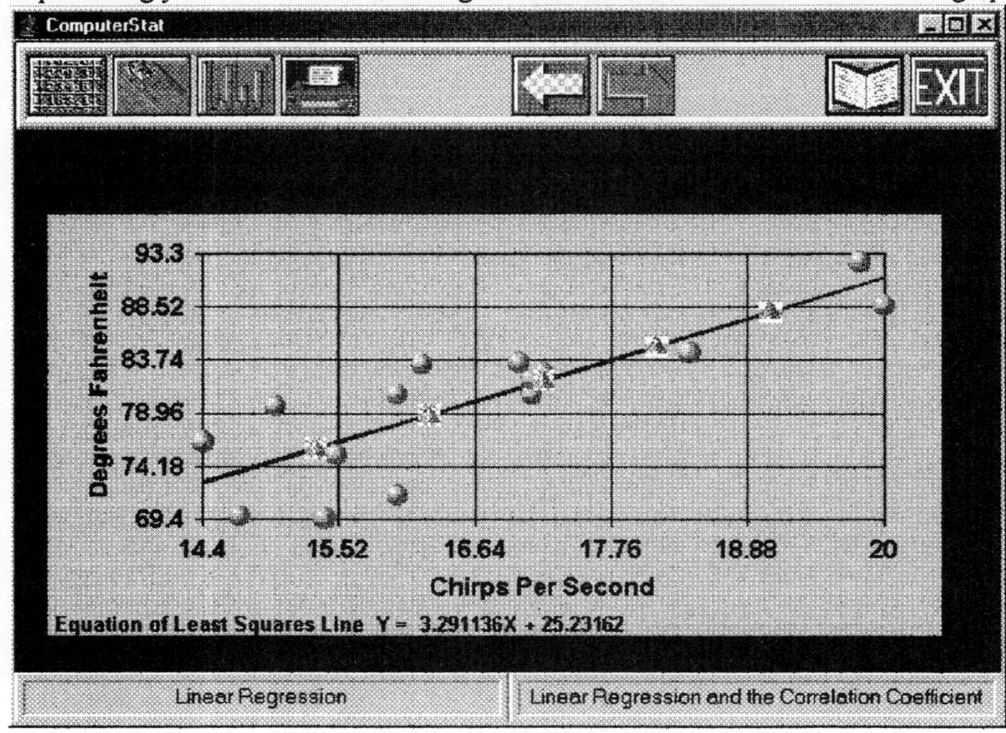

Click on the previous button (left arrow) twice, and then select the prediction option. For x = 18 and a 90% confidence interval for the prediction, we see

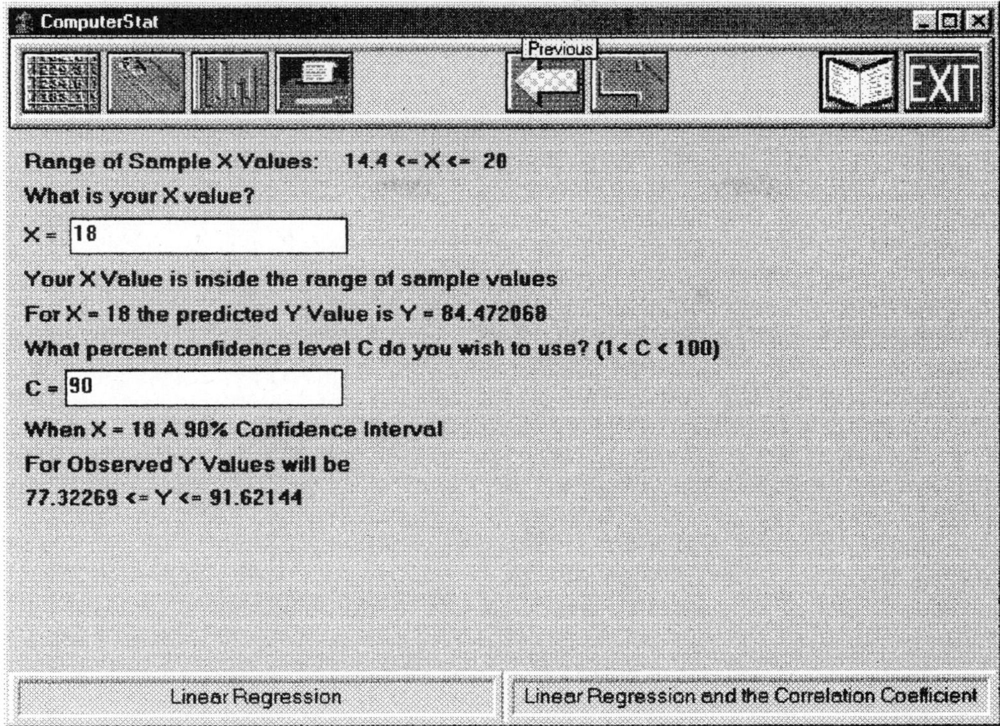

Clicking on the printer will print your results. Click on the previous button and choose the option to test the correlation coefficient.

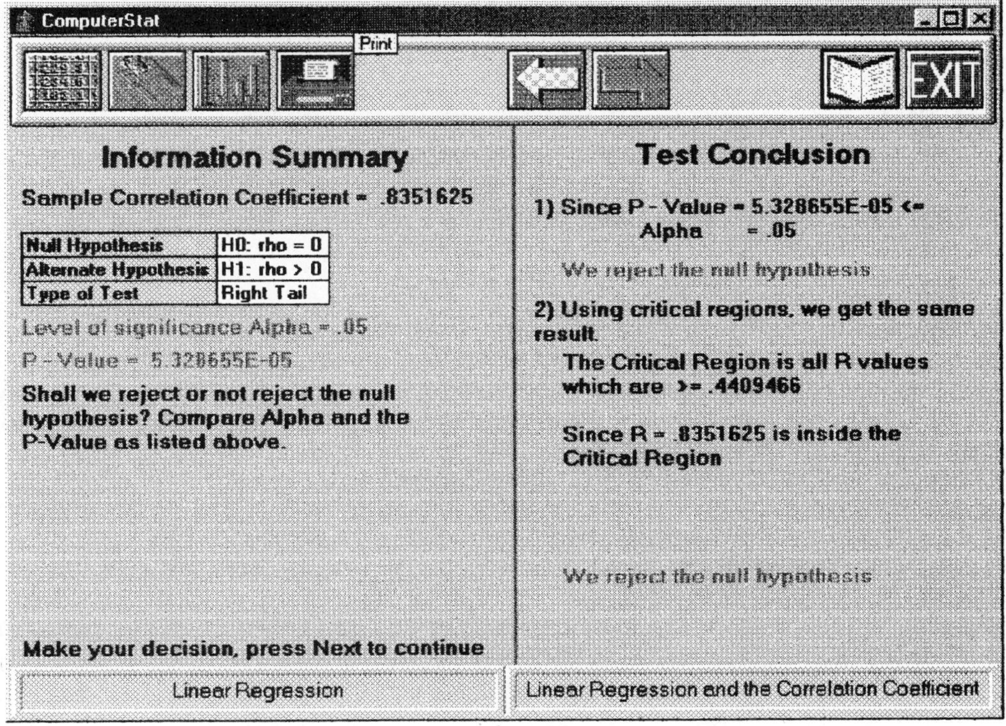

Lab Activities for REGRESSION AND CORRELATION

These activities coordinate with Sections 10.1, 10.2, 10.3, and 10.4 of *Understandable Statistics.*

1. Select Class Demonstration #1: List Price Versus Best Price for a New GMC Pickup Truck. The data are

IN THE FOLLOWING DATA PAIRS (X,Y)

X = LIST PRICE (IN $1000) FOR A GMC PICKUP TRUCK
Y = BEST PRICE (IN $1000) FOR A GMC PICKUP TRUCK
SOURCE: CONSUMERS DIGEST, FEBRUARY 1994

(12.400, 11.200)	(14.300, 12.500)	(14.500, 12.700)
(14.900, 13.100)	(16.100, 14.100)	(16.900, 14.800)
(16.500, 14.400)	(15.400, 13.400)	(17.000, 14.900)
(17.900, 15.600)	(18.800, 16.400)	(20.300, 17.700)
(22.400, 19.600)	(19.400, 16.900)	(15.500, 14.000)
(16.700, 14.600)	(17.300, 15.100)	(18.400, 16.100)
(19.200, 16.800)	(17.400, 15.200)	(19.500, 17.000)
(19.700, 17.200)	(21.200, 18.600)	

a) Look at the scatter plot. Do you think linear regression is appropriate for this data?
b) Look at the information summary sheet and write down the equation of the least squares line, the value of the correlation coefficient, the value of the coefficient of determination, and the value of the standard error of estimate.
c) If the list price of a pickup truck is $20,000 what is the predicted best price? (Note: Be sure to change 20,000 to 20 thousand since the data have been entered in thousands.) Find a 90% confidence interval for the prediction. Would you be surprised if the best price you could get is $19.500?
d) Test the correlation coefficient. At the 5% level of significance, is there evidence of a positive correlation between list and best price?

2. Merchandise loss due to shoplifting, damage, and other causes is called shrinkage. Shrinkage is a major concern to retailers. The managers of H.R. Merchandise thinks that there is a relationship between shrinkage and number of clerks on duty. To explore this relationship, a random sample of 7 weeks was selected. During each week the staffing level of sales clerks was kept constant and the dollar value of the shrinkage was recorded. The results follow.

Number of clerks, X	10	12	11	15	9	13	8
Shrinkage, Y(in $hundreds):	19	15	20	9	25	12	31

a) Select the option to enter your own data. Enter the data. Look at the scatter plot. Does a linear regression model seem appropriate for this model?

Lab Activities for REGRESSION AND CORRELATION continued

b) Look at the information summary sheet and write down the equation of the least squares line, the value of the correlation coefficient, the value of the coefficient of determination, and the value of the standard error of estimate.

c) Does it makes sense to extrapolate beyond the data? Why or why not? What does the model predict for shrinkage if there is a staff of 40 clerks? What about for a staff of 1 clerk? Can you get confidence intervals for these predictions?

d) What does the model predict for a staffing level of 12 clerks? Find a 90% confidence interval for shrinkage.

e) Test the correlation coefficient. Does the sample provide evidence of a negative correlation between staffing level and shrinkage?

MULTIPLE REGRESSION
(Section 10.5 of *Understandable Statistics*)

Main menu selection: REGRESSION AND CORRELATION

Sub-menu selection: MULTIPLE REGRESSION

I. Description of the Program

This program performs a number of functions related to multiple regression. The user inputs a random sample of data points (X_1, X_2, ... , X_7) with up to seven variables. The computer then returns a sample data summary including the mean, standard deviation, and coefficient of variation for each of the variables. Next is a display showing the correlation coefficient between each variable. Following is a display showing a table of the coefficients of determination between each pair of variables. The user then selects which variables will be explanatory variables, and which one will be the response variable. The computer generates the coefficients of the least squares equation and displays statistical information about this equation. Finally the user can predict the value of the response variable for specified values of the explanatory variables and can obtain confidence intervals for the prediction.

Input: Data entry options

Option 1: Select a class demonstration

<<<<< OPTIONS AVAILABLE >>>>>

1) CLASS DEMONSTRATION #1: THUNDER BASIN ANTELOPE STUDY
2) CLASS DEMONSTRATION #2: U. S. ECONOMIC DATA
3) CLASS DEMONSTRATION #3: SYSTOLIC BLOOD PRESSURE DATA (SECTION 10.5 PROBLEM #3)
4) CLASS DEMONSTRATION #4: TEST SCORES FOR GENERAL PSY. (SECTION 10.5 PROBLEM #4)
5) CLASS DEMONSTRATION #5: HOLLYWOOD MOVIES DATA (SECTION 10.5 PROBLEM #5)
6) CLASS DEMONSTRATION #6: ALL GREENS FRANCHISE DATA (SECTION 10.5 PROBLEM #6)

Option 2: Enter your own data with data correction possible. N is the number of variables ($N \leq 7$) and M is the number of data points ($M \leq 20$)

Number of explanatory variables, N1 (from 1 to N - 1)

Identify the explanatory variables by number
Identify the response variable by number
Values of the explanatory variables for which you desire to predict the response variable
Confidence level C of confidence interval for response value.

Output: DISPLAY #1
Number of data points
Mean, standard deviation, and coefficient of variation for each variable

Table showing correlation coefficient between the variables

DISPLAY #2
Coefficient of determination between the variables

DISPLAY #3
List of selected explanatory variables and response variables.
Coefficient of multiple determination
Coefficients of the regression equation with standard error
Degrees of freedom for the t distribution

For given values of the explanatory variable, the predicted value of the response variable and a C% confidence interval for the predicted response variable.

II. Sample run

Bowman Brothers is a large sporting goods store in Denver that has a giant ski sale every year during the month of October. The chief executive officer at Bowman Brothers is studying the following variables regarding the ski sale.

X_1 = Total dollar receipts from October ski sale
X_2 = Total dollar amount spent advertising ski sale on local TV
X_3 = Total dollar amount spent advertising ski sale on local radio
X_4 = Total dollar amount spent advertising ski sale in Denver newspapers

Data for the past eight years is shown below (in thousands of dollars)

Year	1	2	3	4	5	6	7	8
X 1	751	768	801	832	775	718	739	780
X 2	19	23	27	32	25	18	20	24
X 3	14	17	20	24	19	9	10	19
X 4	11	15	16	18	12	5	7	14

a) Use MULTIPLE REGRESSION. Select the enter your own data option. The number of data points is M = 8 with number of variables N = 4. Enter the data points as directed.

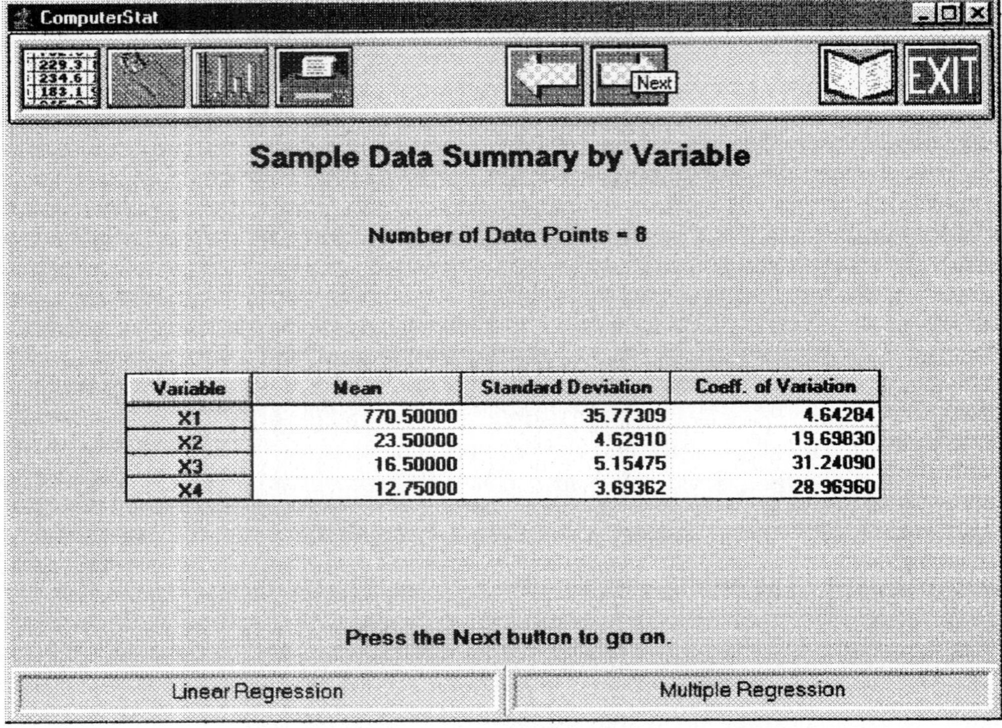

Variable	Mean	Standard Deviation	Coeff. of Variation
X1	770.50000	35.77309	4.64284
X2	23.50000	4.62910	19.69830
X3	16.50000	5.15475	31.24090
X4	12.75000	3.69362	28.96960

Click on the next button to see the correlation coefficients between the variables.

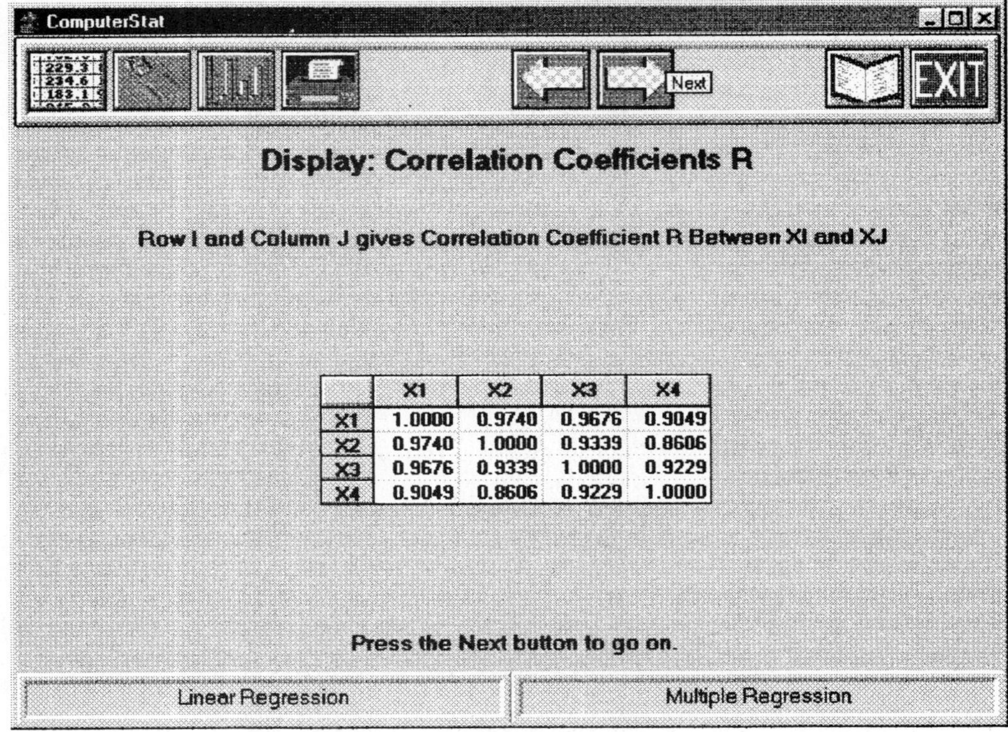

	X1	X2	X3	X4
X1	1.0000	0.9740	0.9676	0.9049
X2	0.9740	1.0000	0.9339	0.8606
X3	0.9676	0.9339	1.0000	0.9229
X4	0.9049	0.8606	0.9229	1.0000

Click on the next button to see the Coefficients of Determination between variables.

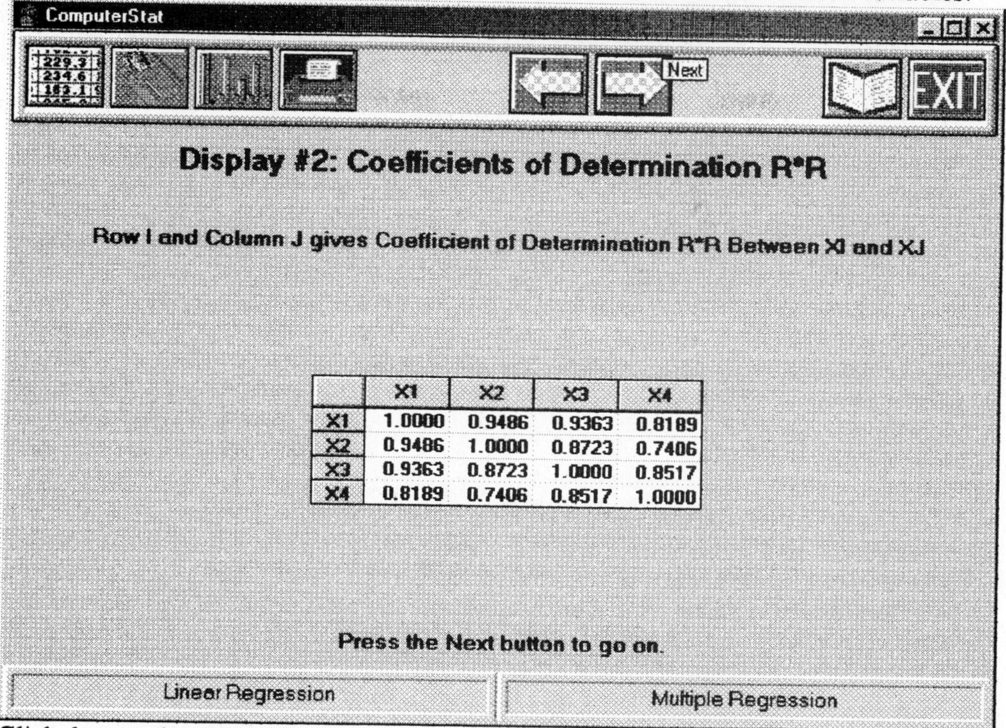

	X1	X2	X3	X4
X1	1.0000	0.9486	0.9363	0.8189
X2	0.9486	1.0000	0.8723	0.7406
X3	0.9363	0.8723	1.0000	0.8517
X4	0.8189	0.7406	0.8517	1.0000

Click the next button to establish a linear regression model. You click on buttons to designate the explanatory variables and the response variable of the model.

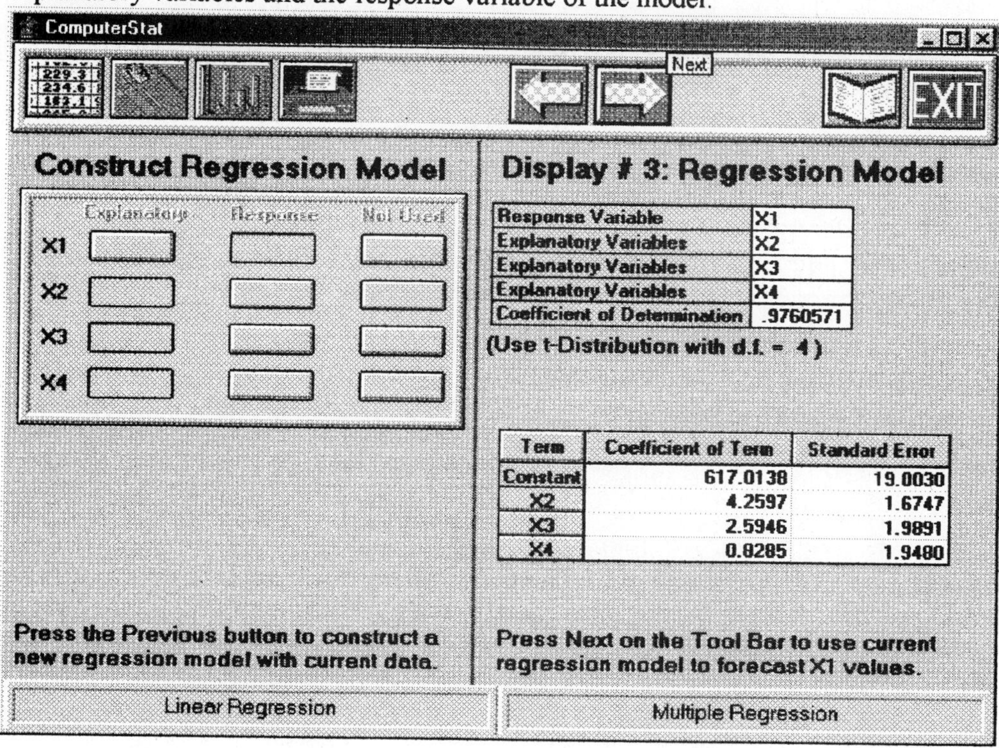

Response Variable	X1
Explanatory Variables	X2
Explanatory Variables	X3
Explanatory Variables	X4
Coefficient of Determination	.9760571

Term	Coefficient of Term	Standard Error
Constant	617.0138	19.0030
X2	4.2597	1.6747
X3	2.5946	1.9891
X4	0.8285	1.9480

To forecast values of the response variable using values of the explanatory variable, click on the next button. Enter the values of the explanatory variables. Click on the next button. After getting the prediction value and entering the confidence level for prediction, click on the next button again to see the confidence interval.

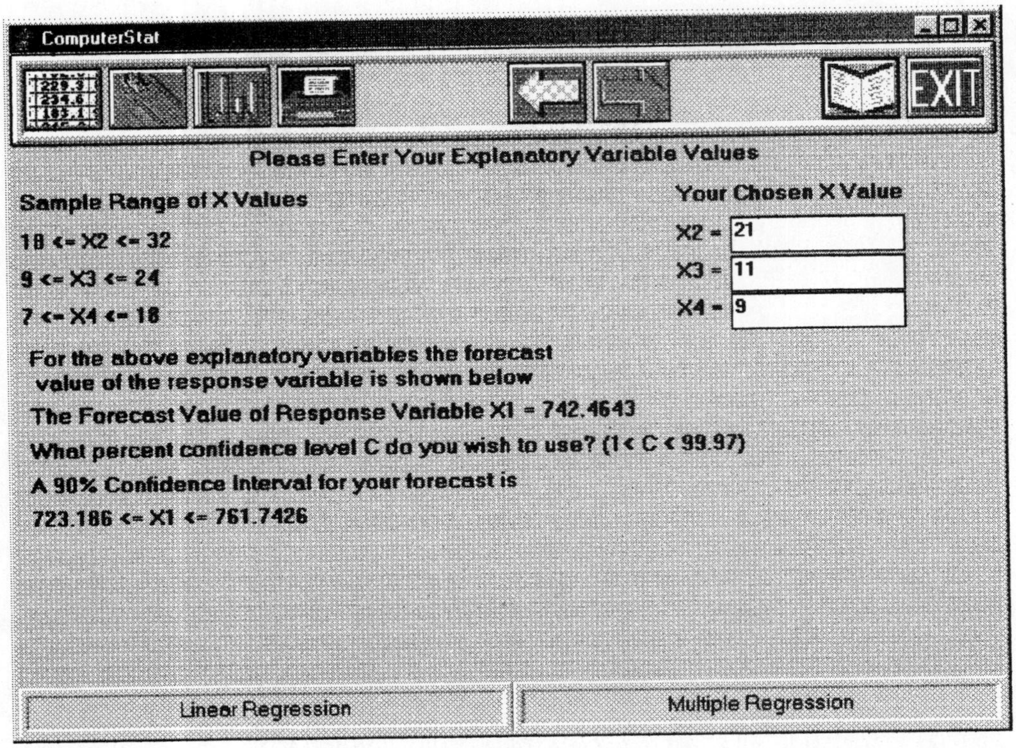

ComputerStat

EXIT

Please Enter Your Explanatory Variable Values

Sample Range of X Values

18 <= X2 <= 32

9 <= X3 <= 24

7 <= X4 <= 18

Your Chosen X Value

X2 = 21

X3 = 11

X4 = 9

For the above explanatory variables the forecast value of the response variable is shown below

The Forecast Value of Response Variable X1 = 742.4643

What percent confidence level C do you wish to use? (1< C < 99.97)

A 90% Confidence Interval for your forecast is

723.186 <= X1 <= 761.7426

Linear Regression | Multiple Regression

Lab Activities for MULTIPLE REGRESSION

These activities coordinate with Section 10.5 of *Understandable Statistics.*

Use the class demonstration files to do the problems in Section 10.5 of *Understandable Statistics.* Also see the section Using Technology at the end of Chapter 10. Do the Multiple Regression Case Study using class demonstration #2.

CHAPTER 11 CHI SQUARE AND F DISTRIBUTIONS

CHI SQUARE TEST FOR INDEPENDENCE
(Section 11.1 of *Understandable Statistics*)

Main menu selection: HYPOTHESIS TESTING

Sub-menu selection: CHI SQUARE TEST FOR INDEPENDENCE

I. Description of the Program

This program uses the Chi Square distribution to test for statistical independence of variables. The hypotheses are

H_0: The variables are independent
H_1: The variables are not independent

The computer requests the user to input the number of rows (R) and the number of columns (C) in the contingency table, the observed frequencies for each cell, and the level of significance alpha for the test. The type of test will always be a right tail test on a chi square distribution with degrees of freedom d.f. = (R - 1)(C - 1).

The expected frequency for each cell in the contingency table is calculated. If the expected frequency in any cell is less than five, a message stating that the sample size is too small will be given and new data will be requested. The output includes a completed contingency table with the observed and expected frequencies. The sample chi square statistic is given as well as the corresponding P-value. The test is concluded by comparing the P-value to the level of significance.

Input: Data Entry Options

Option 1: Select a Class Demonstration

<<<<< OPTIONS AVAILABLE >>>>>

1) CLASS DEMONSTRATION #1: SALARY VERSUS JOB SATISFACTION (SECTION 11.1, PROBLEM #1)
2) CLASS DEMONSTRATION #2: ANXIETY LEVEL VERSUS NEED TO SUCCEED (SEC. 11.1, PROBLEM #2)
3) CLASS DEMONSTRATION #3: AGE VERSES MOVIE PREFERENCE (SECTION 11.1, PROBLEM #3)
4) CLASS DEMONSTRATION #4: NUMBER OF LIBRARY CONTRIBUTORS VERSUS ETHNIC GROUP (SECTION 11.1, PROBLEM #6)

Option 2: Enter your own data with data correction option
Number of rows in the contingency table, R; R between 2 and 10 inclusive
Number of columns in the contingency table, C; C between 2 and 10 inclusive

Level of significance, alpha

Output: Completed contingency table with the row and column number of each cell, the observed frequency in each cell, and the expected frequency of each cell.

Information summary showing the null hypothesis, alternate hypothesis, total sample size, number of rows in the contingency table, number of columns in the contingency table, level of significance, sample chi square value, P-value of the sample statistic.

Test conclusion based on comparison of the P-value to the α.

II. Sample Run

A computer programming aptitude test has been developed for high school seniors. The test designers claim that scores on the test are independent of the type of school the student attends: rural, suburban, urban. A study involving a random sample of students from each of these types of institutions yielded the following information where aptitude scores range from 200-500 with 500 indicating the greatest aptitude and 200 the least. The entry in each cell is the observed number of students making the indicated score on the test.

	School Type		
Score	Rural	Suburban	Urban
200 - 299	1 33	2 65	3 82
300 - 399	4 45	5 79	6 95
400 - 500	7 21	8 47	9 63

Using the program CHI SQUARE TEST OF INDEPENDENCE, test the claim that the aptitude test scores are independent of the type of school attended at the 0.05 level of significance.

We have R = 3 rows and C = 3 columns. Enter these numbers when requested. The level of significance is alpha = 0.05. Enter the data of the contingency table by rows. The results are

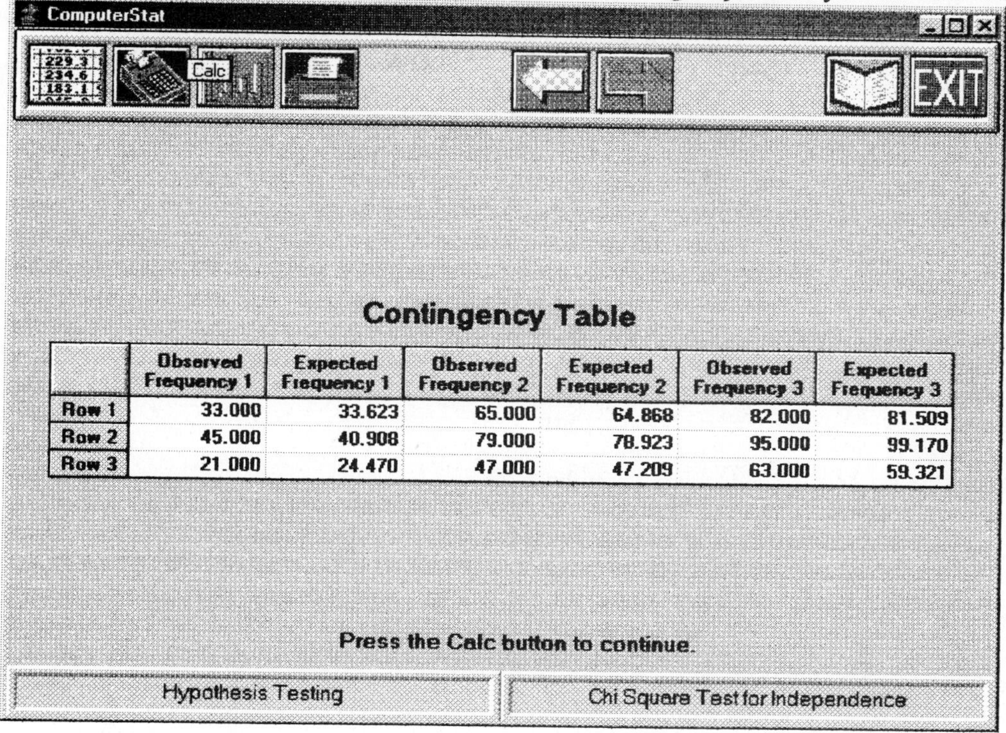
ComputerStat
Calc
EXIT
Contingency Table
Press the Calc button to continue.
Hypothesis Testing
Chi Square Test for Independence

	Observed Frequency 1	Expected Frequency 1	Observed Frequency 2	Expected Frequency 2	Observed Frequency 3	Expected Frequency 3
Row 1	33.000	33.623	65.000	64.868	82.000	81.509
Row 2	45.000	40.908	79.000	78.923	95.000	99.170
Row 3	21.000	24.470	47.000	47.209	63.000	59.321

Click on the calculator button to conclude the test.

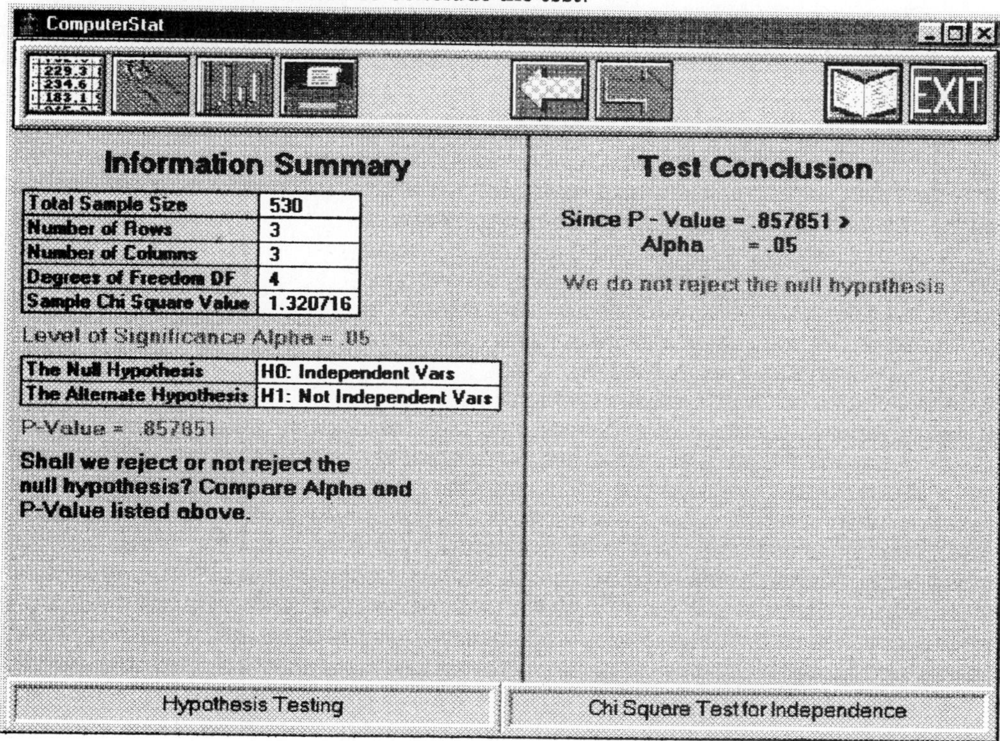
ComputerStat
EXIT
Information Summary
Level of Significance Alpha = .05
P-Value = .857851
Shall we reject or not reject the null hypothesis? Compare Alpha and P-Value listed above.
Test Conclusion
Since P-Value = .857851 >
Alpha = .05
We do not reject the null hypothesis
Hypothesis Testing
Chi Square Test for Independence

Total Sample Size	530
Number of Rows	3
Number of Columns	3
Degrees of Freedom DF	4
Sample Chi Square Value	1.320716

The Null Hypothesis	H0: Independent Vars
The Alternate Hypothesis	H1: Not Independent Vars

Lab Activities for CHI SQUARE TEST OF INDEPENDENCE

These activities coordinate with Section 11.1 of *Understandable Statistics.*

1. We Care Auto Insurance had its staff of actuaries conduct a study to see if vehicle type and loss claim are independent. A random sample of auto claims over the 1st six months give the information in the contingency table.

Total Loss Claims per Year per Vehicle

Type of vehicle	\$0-999	\$1000-2999	\$3000-5900	\$6000+
Sports car	20	10	16	8
Truck	16	25	33	9
Family Sedan	40	68	17	7
Compact	52	73	48	12

Test the claim that car type and loss claim are independent. Use $\alpha = 0.05$.

2. An educational specialist is interested in comparing three methods of instruction:

S.L.- standard lecture with discussion
T.V.- video taped lectures with no discussion
I.M.- individualized method with reading assignments and tutoring, but no lectures.

The specialist conducted a study of these three methods to see if they are independent. A course was taught using each of the three methods and a standard final exam was given at the end. Students were put into the different method sections at random. The course type and test results are shown in the contingency table.

Final Exam Score

Course Type	below 60	60-69	70-79	80-89	90-100
S.L.	10	4	70	31	25
T.V.	8	3	62	27	23
I.M.	7	2	58	25	22

Test the claim that the instruction method and final exam test scores are independent using $\alpha = 0.01$.

3. Select one of the class demonstrations and use *ComputerStat* to do the corresponding Section 11.1 problem of *Understandable Statistics.*

CHI SQUARE GOODNESS OF FIT

(Section 11.2 of *Understandable Statistics*)

Main menu selection: HYPOTHESIS TESTING

Sub-menu selection: CHI SQUARE GOODNESS OF FIT

I. Description of the Program

This program uses the chi square distribution to test for goodness of fit. The hypotheses are

H_0: The population from which sample measurements are taken fits a given distribution.
H_1: The population from which sample measurements are taken does not fit the given distribution.

The user inputs the number of items (that is, categories) in the sample distribution, the total sample size, the observed sample frequency for each item, and the expected percent of total sample size for each item. The level of significance is also requested.

The type of test will always be a right tail test on a chi square distribution with degrees of freedom d.f. = (Number of items - 1). The computer will calculate the expected frequency for each item. If the expected frequency for any item is less than five, a message stating that the sample size is too small will be given.

The test conclusion is based on the P-value of the sample chi square statistic.

Input: Data entry options
Option 1: Use a class demonstration

<<<<< OPTIONS AVAILABLE >>>>>

1) CLASS DEMONSTRATION #1: LIBRARY CIRCULATION DISTRIBUTION
(SECTION 11.2, PROBLEM #6)
2) CLASS DEMONSTRATION #2: ELECTRICAL POWER DISTRIBUTION
(SECTION 11.2, PROBLEM #7)
3) CLASS DEMONSTRATION #3: HOSPITAL PATIENT DISTRIBUTION
(SECTION 11.2, PROBLEM #8)
4) CLASS DEMONSTRATION #4: ETHNIC DISTRIBUTION
(SECTION 11.2, PROBLEM #9)

Option 2: Enter your own data
Number of items in distribution; value between 2 and 50 inclusive
Observed frequency for each item, O(I)
Expected percent of total sample size per item, E1(I)

Output: Expected frequency per item; E(I)
Information summary
Sample chi square value
P-value

II. Sample Run

Wenland College just changed the length of its summer session from 10 weeks to 8 weeks. The director of the self pace individualized instruction center wants to see if the grade distribution for courses offered in the center has changed with the shortened summer session. Past records given the grade distribution for courses conducted in a 10 week session. from thee figures, the expected percent of the sample size can be computed for each item. A random sample of 216 students showed the grade distribution for the 8 week summer session.

Item Grade	Observed Frequency from sample (8 weeks)	Expected Frequency % (10 week session)
A	28	15
B	73	40
C	40	15
D	8	3
Incomplete	49	18
Withdrawal	18	9

Use **CHI SQUARE GOODNESS OF FIT** to see if the grade distribution for the 8 week summer session is different from that of the 10 week session at the 0.05 level of significance.

The sample size is n = 216. The number of items in the sample distribution is 6. Enter that value when requested. The level of significance is 0.05. Next enter the observed frequencies and expected percent of total sample size. The computer immediately computes the expected frequency for that item. A data summary is provided, followed by the information summary.

After entering the data, click on the calculator key.

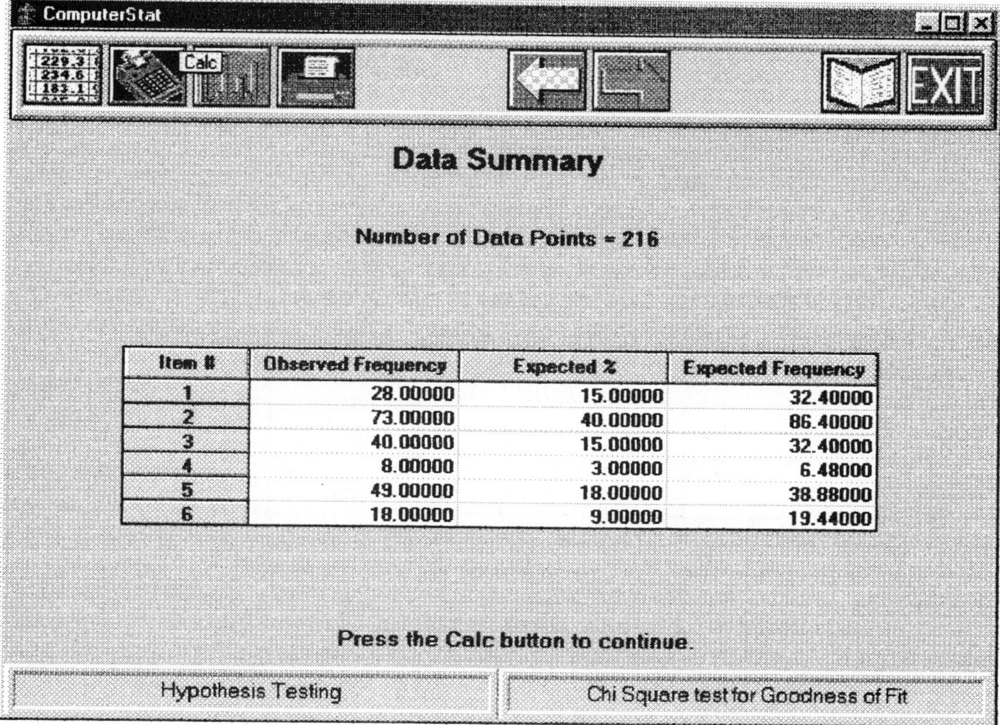
ComputerStat
229.3
234.6
183.1
Calc
EXIT
Data Summary
Number of Data Points = 216
Press the Calc button to continue.
Hypothesis Testing
Chi Square test for Goodness of Fit

Item #	Observed Frequency	Expected %	Expected Frequency
1	28.00000	15.00000	32.40000
2	73.00000	40.00000	86.40000
3	40.00000	15.00000	32.40000
4	8.00000	3.00000	6.48000
5	49.00000	18.00000	38.88000
6	18.00000	9.00000	19.44000

Click on the calculator key again to conclude the test.

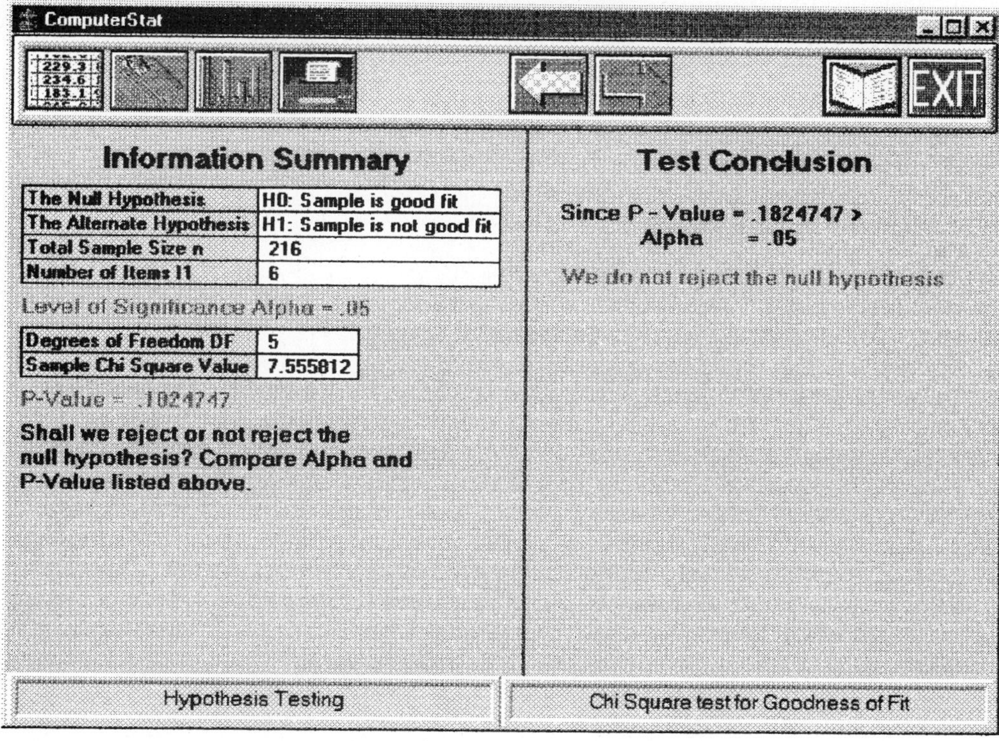
ComputerStat
229.3
234.6
183.1
EXIT
Information Summary
Level of Significance Alpha = .05
P-Value = .1824747
Shall we reject or not reject the
null hypothesis? Compare Alpha and
P-Value listed above.
Test Conclusion
Since P - Value = .1824747 >
Alpha = .05
We do not reject the null hypothesis
Hypothesis Testing
Chi Square test for Goodness of Fit

The Null Hypothesis	H0: Sample is good fit
The Alternate Hypothesis	H1: Sample is not good fit
Total Sample Size n	216
Number of Items I1	6

Degrees of Freedom DF	5
Sample Chi Square Value	7.555812

Lab Activities for CHI SQUARE TEST OF GOODNESS OF FIT

1. Market Survey did a study of moderate priced restaurants (diner for less than $15) according to type. Five years ago they had done a similar study in the Midwest. A new study is being conducted in the same area of the country to determine if the distribution of restaurants has changed. A random sample of 129 restaurants in the Midwest was used.

Item	Observed Frequency	Expected Frequency %
American food	51	47
Chinese food	14	6
Mexican food	15	10
Italian food	27	22
Other ethnic food	14	10
Salad bar	8	5

   At the 5% level of significance, test the claim that there has been a shift in the distribution of restaurants according to food type.

2. Select one of the class demonstrations and use *ComputerStat* to do the corresponding Section 11.2 problem of *Understandable Statistics.*

ANOVA
(Section 11.4 of *Understandable Statistics*)

Main menu selection: HYPOTHESIS TESTING

Sub-menu selection: ANALYSIS OF VARIANCE

I. Description of the Program

This program is for single factor analysis of variance (also called ANOVA or one-way ANOVA). The user inputs the number of groups, the number of subjects in each group, and the measurement for each of the subjects. The user also inputs the level of significance, alpha for the test.

The hypotheses are:

H_0: All group population means are equal
H_1: Not all group population means are equal

The type of test is a right tail on the F distribution with appropriate degrees of freedom in numerator and denominator.

The computer will print an information summary of all important data and calculations. The P-value of the sample F-ratio is compared to the level of significance to conclude the test.

Input: Data entry options

Option 1: Select a Class Demonstration

<<<<< OPTIONS AVAILABLE >>>>>

1) CLASS DEMONSTRATION #1: PATTERN RECOGNITION EXPERIMENT (SEC. 11.4, GUIDED EXERCISE 10)
2) CLASS DEMONSTRATION #2: RIVER ECOLOGY EXPERIMENT (SECTION 11.4, PROBLEM #2)
3) CLASS DEMONSTRATION #3: SALES IN BRANCH OFFICES (SECTION 11.4, PROBLEM #3)
4) CLASS DEMONSTRATION #4: SOCIOLOGY STUDY IN NEW YORK CITY (SECTION 11.4, PROBLEM #5)

Option 2: Enter your own data

Number of groups (treatments), K; K between 2 and 10 inclusive
Number of subjects in the Ith group, N1(I); between 2 and 30 inclusive
Measurement for the Jth subject in the Ith group XI(J)

Level of significance

Output: Information summary with

Null and alternate hypotheses
Type of test - right tail on F-distribution
Number of data groups
Total sample size

Sum of squares SS between groups, within groups, total
Degrees of freedom DF between groups, within groups, total
Mean Square MS between groups and within groups
Sample F-ratio
Level of significance
P-value

Test conclusion based on comparison of P-value to level of significance

II. Sample run

A psychologist has developed a series of tests to measure depression level. The composite scores range from 50 to 100 with 100 representing the most severe depression level. This measuring device was used in a study of treatments for depression. A random sample of 12 patients with approximately the same depression level as measured by the tests was divided into 3 different treatment groups. Then, one month after treatment was competed, the depression level of each patient was again evaluated using the series of tests. The after treatment depression levels are given

Treatment 1: 70 65 82 83 71
Treatment 2: 75 62 81
Treatment 3: 77 60 80 75

Use the program ANALYSIS OF VARIANCE to test the claim that the population means are all the same at the 5% level of significance.

For this problem, use 3 for the number of groups. There are 5 subjects in group 1, 3 subjects in group 2, and 4 subjects in group 3. Enter the data in the table provided. Be sure all the data for treatment 1 is in the first column, etc. The level of significance is 0.05.

The information summary and test conclusion are shown on the next page.

The summary statistics and test conclusion follow.

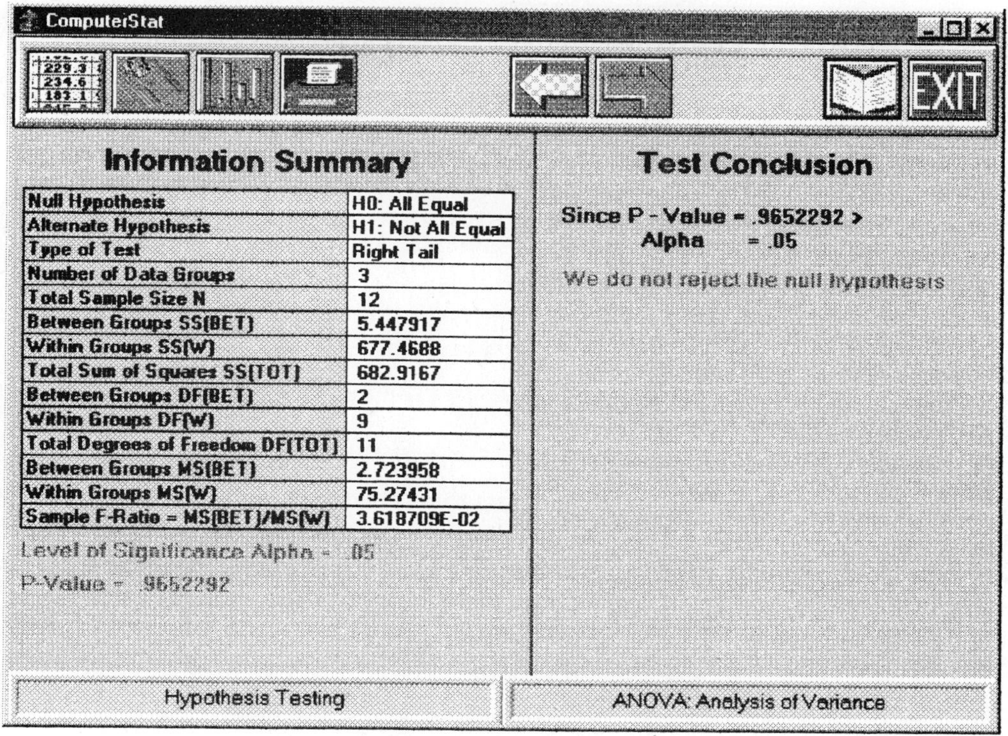

Lab Activities for ANALYSIS OF VARIANCE

These activities coordinate with section 11.4 of *Understandable Statistics.*

1 A random sample of 20 overweight adults were randomly divided into 4 groups. Each group was given a different diet plan, and the weight loss for each individual after 3 months follows:

Plan 1: 18 10 20 25 17
Plan 2: 28 12 22 17 16
Plan 3: 16 20 24 8 17
Plan 4: 14 17 18 5 16

Test the claim that the population mean weight loss is the same for the four diet plans at the 5% level of significance.

2. A psychologist is studying the time it takes rats to respond to stimuli after being given doses of different tranquilizing drugs. A random sample of 18 rats were divided into 3 groups. Each group was given a different drug. The response time to stimuli was measured (in seconds). The results follow.

Drug A 3.1 2.5 2.2 1.5 0.7 2.4
Drug B 4.2 2.5 1.7 3.5 1.2 3.1
Drug C 3.3 2.6 1.7 3.9 2.8 3.5

Test the claim that the population mean response times for the three drugs is the same at the 5% level of significance.

3. A research group is testing various chemical combinations designed to neutralize and buffer the effects of acid rain on lakes. A random sample of 18 lakes of similar size in the same region have all been affected in the same way by acid rain. The lakes are divided into four groups and each group of lakes is sprayed with a different chemical combination. An acidity index is then taken after treatment. The index ranges from 60 to 100 with 100 indicating the greatest acid rain pollution. The results follow.

Combination I 63 55 72 81 75
Combination II 78 56 75 73 82
Combination III 59 72 77 60
Combination IV 72 81 66 71

Test the claim that the population mean acidity index after each of the four treatments is the same at the 0.01 level of significance.

PART III

MINITAB

FOR

UNDERSTANDABLE STATISTICS
SIXTH EDITION

OR

UNDERSTANDING BASIC STATISTICS

CHAPTER 1 GETTING STARTED

MINITAB, Student Edition, Release 12

In this chapter you will find
a) general information about MINITAB.
b) general directions for using the Windows style pull-down menus
c) general instructions for choosing values for dialog boxes
d) how to enter data
d) other general commands.

General Information

MINITAB is a command driven software package with more than 200 commands available. However, in Windows versions of MINITAB, menu options and dialog boxes can be used to generate the appropriate commands. After using the menu options and dialog boxes, the actual commands are shown in the session window along with the output of the desired process. Data are stored and processed in a table with rows and column. Such a table is similar to a spreadsheet, It is called a worksheet. Unlike electronic spreadsheets, a MINITAB worksheet can contain only numbers and text, not formulas. Constants are also stored in the worksheet, but are not visible.

MINITAB will accept words typed in upper or lower case letters as well as a combination of the two. Comments elaborating on the commands may be included. In this *Guide* we will follow the convention of typing the essential parts of a command in upper case letters and optional comments in lower case letters.

COMMAND with comments

Note that only the *first four letters* of a command are essential. However, we usually give the entire command name in examples.

Numbers must be typed *without* commas. Be sure you type the number zero instead of the letter "O" and that you use the number one (1) instead of a lower case letter "l". Exponential notation is also acceptable. For instance

127.5   1.257E2   1.257E+2

are all acceptable in MINITAB and have the same value.

The MINITAB *worksheet* contains columns, rows, and constants.

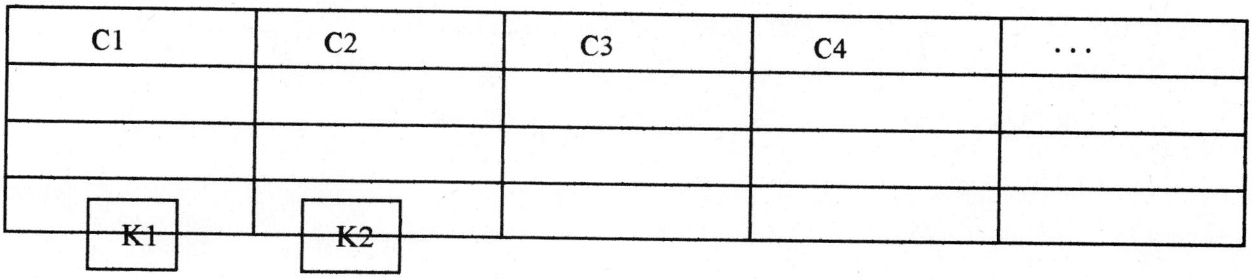

C1	C2	C3	C4	. . .

K1   K2

The columns are designated by the letter **C** followed by a number

C1, **C2**, **C3** designate columns 1, 2, and 3

Constants require the letter **K**, and may be followed by a number if there are several constants.

K1 K2 designate constant 1 and constant 2 respectively

Starting and Ending MINITAB

The steps you use to start MINITAB will differ according to the computer equipment you are using. You will need to get specific instructions for your installation from your professor or computer lab manager. Use this space to record the details of logging onto your system and accessing MINITAB. For Windows versions, you generally click on the MINITAB icon to begin the program.

The first screen will look similar to

MINITAB®
For Windows

The Student Edition of MINITAB for Windows 95 / Windows NT
Based on MINITAB Release 12

This software is licensed for use by:
cbrase
acc
WINSE12.11

Minitab Inc.
http://www.minitab.com

Copyright [C] 1998, Minitab Inc.

After a pause, a split screen appears. The session window is at the top, and the data window is at the bottom.

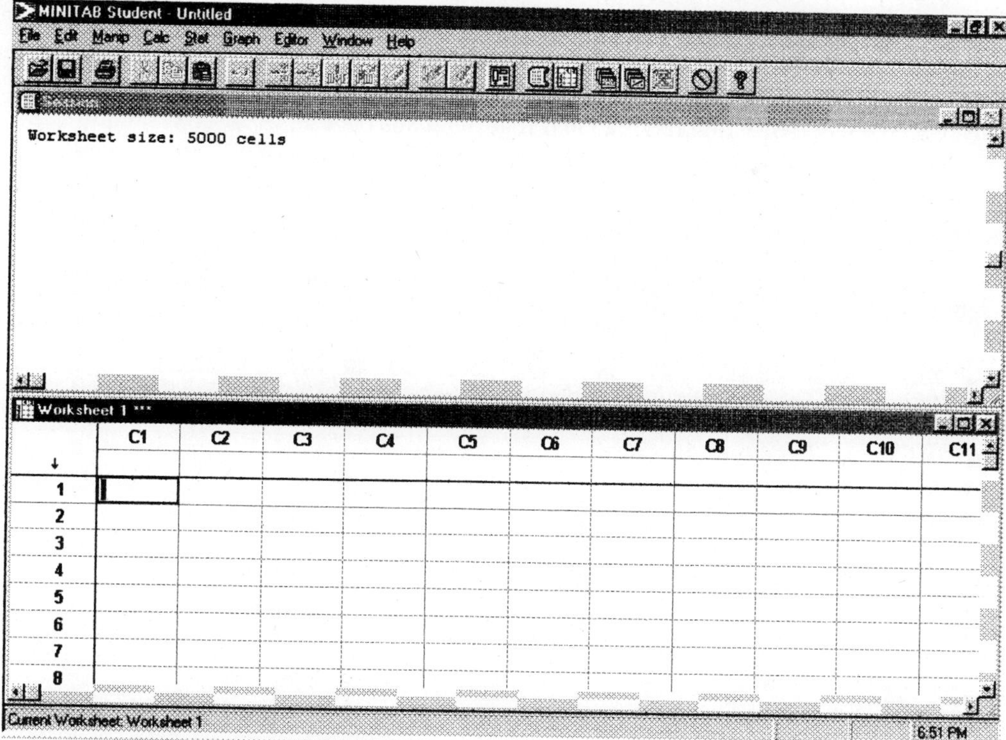

Notice the main menu items are

File Edit Manip Calc Stat Graph Editor Window Help

The toolbar contains icons for frequently used operations.

We enter data in the individual cells in the Data Window. We can enter commands directly in the Session Window, or use the menu options together with dialog boxes to generate the commands.

To end MINITAB: Click on the **File** option. Select **Exit** and click on it or press Enter.
Menu selection summary: **➤File ➤Exit**

Entering Data

One of the first tasks you do when you begin a MINITAB session is to insert data into the worksheet. The easiest way to enter data is to enter it directly into the Data Worksheet. Notice that the active cell is outlined by a heavier box.

To enter a number type it in the active box and then press ENTER or TAB. The data value is entered and the next cell is activated. Data for a specific variable are usually entered by column. Notice that the there is a cell for a column label above row number 1.

To change a data value in a cell, click on the cell, correct the data, and press ENTER or TAB

Example to Create a Worksheet

Open a new worksheet by selection ➤**File** ➤**New**. (Note: You can have a maximum of 5 worksheets open at one time. If necessary, close some)

Let's create a new worksheet that has data in it regarding ads on TV. A random sample of 20 hours of prime time viewing on TV gave information about the number of TV ads in each hour as well as the total time consumed in the hour by ads. We will enter the data into two columns: one column representing the number of ads and the other the time per hour devoted to ads.

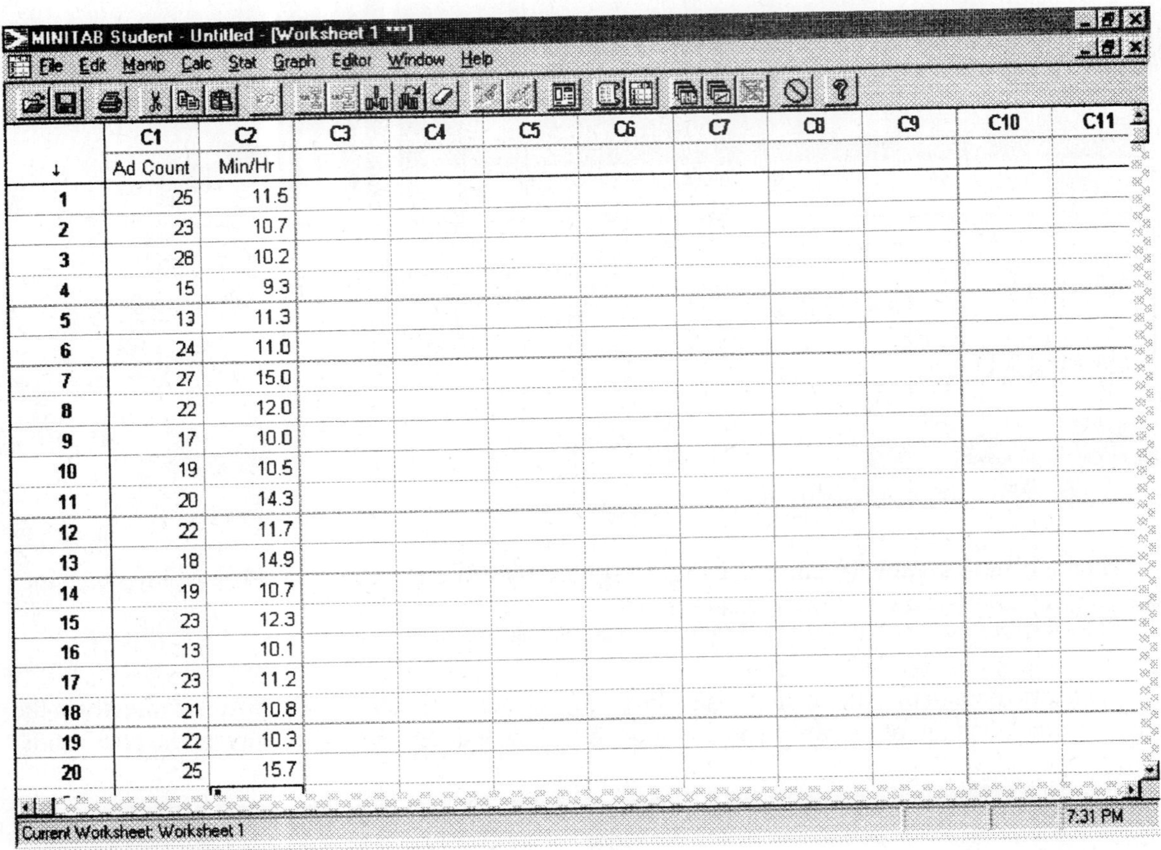

↓	C1 Ad Count	C2 Min/Hr	C3	C4	C5	C6	C7	C8	C9	C10	C11
1	25	11.5									
2	23	10.7									
3	28	10.2									
4	15	9.3									
5	13	11.3									
6	24	11.0									
7	27	15.0									
8	22	12.0									
9	17	10.0									
10	19	10.5									
11	20	14.3									
12	22	11.7									
13	18	14.9									
14	19	10.7									
15	23	12.3									
16	13	10.1									
17	23	11.2									
18	21	10.8									
19	22	10.3									
20	25	15.7									

Notice that we typed a name for each column. Try to keep the name at 8 characters or fewer. Longer names may be truncated in data displays.

To switch between the Data Window and the Session Window, click on the Window menu and select the window you want. The Data Window is called the Worksheet Window.

Working with the Data and Printing the Data

There are several commands for inserting or deleting rows or cells. One way to access these commands is to use the Manip menu option or the Edit menu option..

Click on the Manip menu item or the Edit menu item. You will see these cascading options.

Manip	**Edit**
Subset Worksheet	**Clear Cells**
Split Worksheet	**Delete Cells**
Sort	**Copy Cells**
Rank	**Cut Cells**
Delete rows	**Select All Cells**
Erase variables	**Edit Last Dialogue**
Copy Columns	**Command Line Editor**
Stack/Unstack	**Preferences**
Concatenate	
Code	
Change Data Type	
Display Data	

A useful item is **Change Data Type**. If you accidently typed a letter instead of a number, you have changed the data type to alpha. To change it back to numeric, use **➤Manip➤Change Data Type** and fill in the dialogue box. The same process can be used to change back to alpha.

If you want to see the data displayed in the session window, select **➤Manip ➤Display Data** and select the columns you want to see displayed.

To **print** the worksheet, click the cell in the upper-left corner of the Data window. Press [Shift] + [Ctrl] + [End] to highlight the entire Data window containing all your data. Then click on the printer icon on the toolbar. You can also select **➤File ➤Print Worksheet** from the menus.

Manipulating the Data

You can also do calculations with entire columns. Click on the **Calc** menu item and select **Calculator** (**➤Calc ➤Calculator**). The dialogue box appears

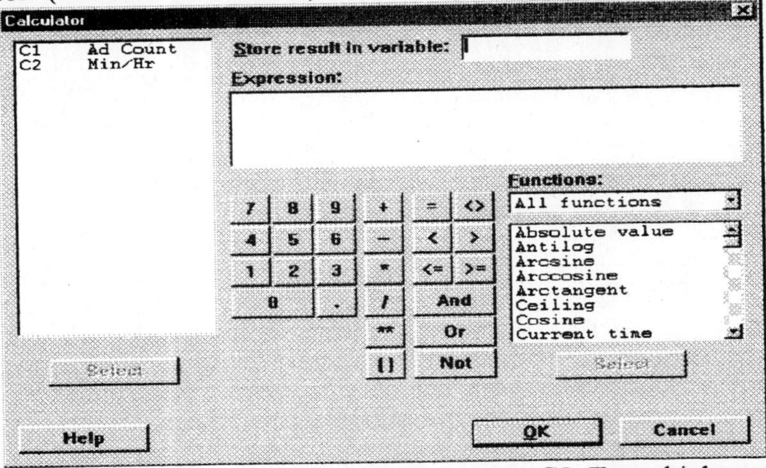

You can store the results in a new column, say C3. To multiply each entry from C1 by 3 and add 4, type 9 , click on the multiply key * on the calculator, type C1, click on the + key on the calculator, type 4. Click on OK. The results of this arithmetic will appear in column C3 of the data sheet.

Saving a Worksheet

Click on the File menu item and select **Save Current Worksheet As**. A dialogue box similar to the following appears. (Select ➤ **File ➤Save Current Worksheet As)**

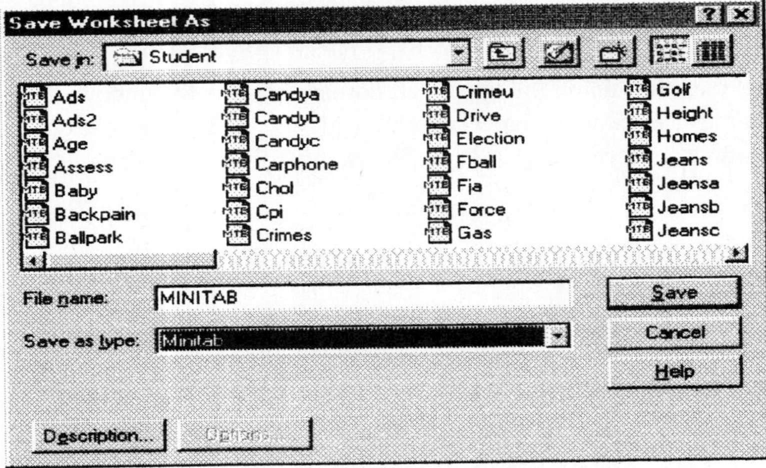

For most computer labs, you will save your file on a 3½ inch floppy. Insert it in the appropriate drive. Scroll down the Save in button until you find 3½ floppy [A}. Then select a file name. In most cases you will save the file as a **Minitab** file. If you change versions of MINITAB or systems, you might select **Minitab portable**.

Example

Let's save the worksheet created in the previous example (information about ads on TV).

If you added column C3 as described under Manipulating data, highlight all the entries of the column and press the Del key. Your worksheet should have only two columns. Use ➤**File** ➤**Save** Current Worksheet As. Insert a diskette in Drive A. Scroll down the Save in box and select 3½ Floppy A. Name the file Ads. Click on [Save]. The worksheet will be saved as Ads.mtw..

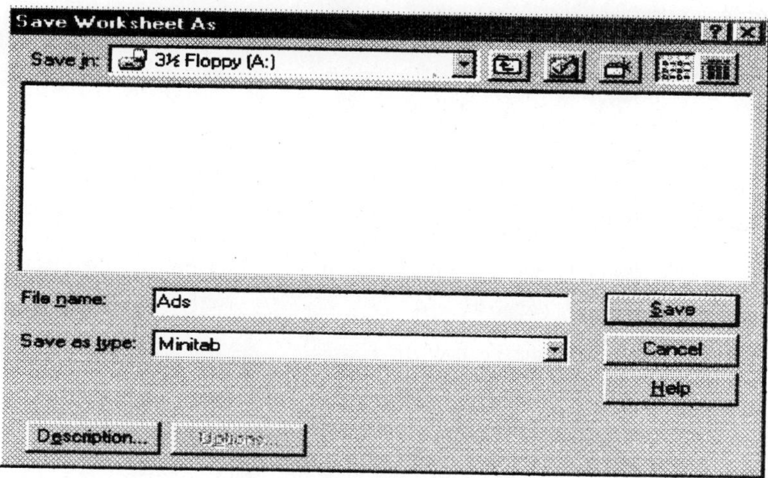

Lab Activities

1. Go to your computer lab (or use your own computer) and learn how to access MINITAB.

2. a) Use the data worksheet to enter the data
   1 3.5 4 10 20 in Column C1
   Enter the data
   3 7 9 8 12 in Column C2
   b) Use ➤**Calc** ➤ **Calculator** to create C3. The data in C3 should be 2*C1 + C2. Check to see that the first entry in C3 is 5. Do the other entries check?
   c) Name C1 First, C2 Second, C3 Result
   c) Save the worksheet as Prob2 on a floppy diskette
   d) Retrieve the worksheet by selecting ➤**File** ➤**Open Worksheet**.
   e) Print the worksheet. Use either the Print Worksheet button or select ➤**File** ➤**Print Worksheet**

Command Summary

Instead of using menu options and dialogue boxes, you can type commands directly into the session window. Notice that you can enter data via the session window with the commands **READ** and **SET** rather than through the data window. The following commands will enable you to open worksheets, enter data, manipulate data, save worksheets, etc. Note: Switch to the Session Window, the menu choice ➤**Editor** ➤**Enable Command Language** allows you to enter commands directly in the Sessions Window and also shows the commands corresponding the to menu choices

HELP gives general information about MINITAB
WINDOWS menu: **Help**

INFO gives the status of the worksheet

STOP ends MINITAB session
WINDOWS menu: **File ➤ Exit**

<u>To Enter Data</u>

READ C...C puts data into designated columns
READ 'filename' C...C reads data from file into columns
SET C puts data into single designated column
SET 'filename' C reads data from file into column
END signals end of data
NAME C = 'name' names column C
WINDOWS menu: You can enter data in rows or columns and name the column in the DATA window. To access the data window select **Window ➤ Data**

RETRIEVE 'filename' retrieves worksheet
WINDOWS menu: **File ➤ Retrieve**

<u>To Edit Data</u>

LET C(K) = K changes the value in row K of column C
INSERT K K C C inserts data between rows K and K into columns C to C
DELETE K K C C deletes data between rows K and K from columns C to C
WINDOWS menu: You can edit data in rows or columns in the DATA window. To access the data window select **Window ➤ Data**

COPY C into C copies column C into column C
USE rows K...K subcommand to copy designated rows
OMIT rows K...K subcommand to omit designated rows
WINDOWS menu: **Manip ➤ Copy Columns**

ERASE E...E erases designated columns or constants
WINDOWS menu: **Manip ➤ Erase Variables**

To Output Data

PRINT E...E prints designated columns or constant
WINDOWS menu: **File ➤ Display Data**

SAVE 'filename' saves current worksheet
PORTABLE subcommand to make worksheet portable
WINDOWS menu: **File ➤ Save Worksheet**
WINDOWS menu: **File ➤ Save Worksheet As...** you may select portable

WRITE 'filename' C...C saves data in ASCII file
WINDOWS menu: **File ➤ Other Files ➤ Export ASCII Data**

Miscellaneous

PAPER prints session
NOPAPER stops printing session
WINDOWS menu: **File ➤ Print Window**

OUTFILE = 'filename' saves session in ASCII file
NOOUTFILE ends OUTFILE
WINDOWS menu: **File ➤ Other Files ➤ Start/Stop Recording**

CHAPTER 2 ORGANIZING DATA

RANDOM SAMPLES

(Section 2.1 Random Samples of *Understandable Statistics*)

In MINITAB you can take random samples from a variety of distributions. We begin with one of the simplest: random samples from a range of consecutive integers under the assumption that each of the integers is equally likely to occur. The basic command RANDOM draws the random sample, and subcommands refer to the distribution being sampled. To sample from a range of equally likely integers, we use the subcommand INTEGER. The menu selection options are.

➤Calculate ➤Random Data➤Integer

Responses for the dialogue box:

Generate ___ rows of data: Enter the sample size
Store in: Enter the column number C# in which you wish to store the sample numbers
Minimum: Enter the minimum integer value of your population
Maximum: Enter the maximum integer value of your population

The random sample numbers are given in the order of occurrence. If you want them in ascending order (so you can quickly check to see if any values are repeated), use the SORT command.

➤Manip➤Sort

Responses for the dialogue box:
Sort columns: Enter the column number C# containing the data you wish to sort
Store sorted column in: Enter the column number C# where you want to store the sorted data. This may the same column number as that containing the original unsorted data
Sort by column: Enter the same column number C# that contains the original data. Leave the rest of the sort by columns options empty

Example

There are 175 students enrolled in a large section of introductory statistics. Draw a random sample of 15 of the students.

We number the students from 1 to 175, so we will be sampling from the integers 1 to 175. We don't want any student repeated, so if our initial sample has repeated values, we will continue to sample until we have 15 distinct students. We sort the data so that we can quickly see if any values are repeated.

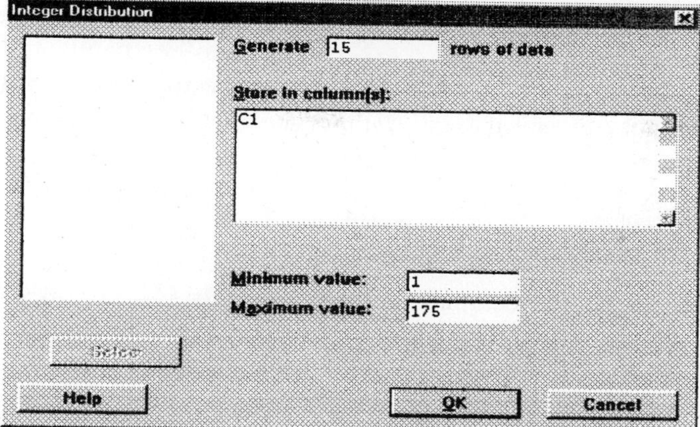

Next sort the data

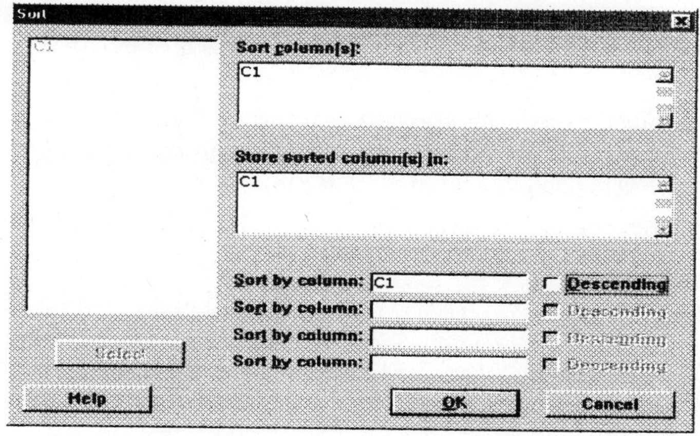

Switch to the Data Window and type the name Sample as the header to C1. To display the data, use the command **➤Manip➤Display Data**. The results are shown.

```
Sample
   8      13      31      33      36      66      74      80      111
   123     138     140     141     158     162
```

We see that no data are repeated. If there had been repetitions, keep sampling until you get 15 distinct values.

Random numbers are also used to simulate activities or outcomes of a random experiment such as tossing a die. Since the six outcomes 1 through 6 are equally likely, we can use the RANDOM command with the INTEGER subcommand to simulate tossing a die any number of times. When outcomes may occur repeatedly it is convenient to tally, count, and give percents of the outcomes. We do this with the TALLY command and appropriate subcommands.

➤Stat ➤Table ➤Tally
Dialogue Box Responses:
Variables: Column number C# or column name containing data
Option to check: ; counts, percents, cumulative counts, cumulative percents.

Example

Use the RANDOM command with INTEGER A = 1 to B = 6 subcommand to simulate 100 tosses of a fair die. Use the TALLY command to give a count and percent of outcomes.

Generate the random sample using the menu selection **➤Random Data➤ Integer**, with generate at 100, min at 1 and max at 6. Then use **➤Stat ➤Table ➤Tally** with counts and percents checked

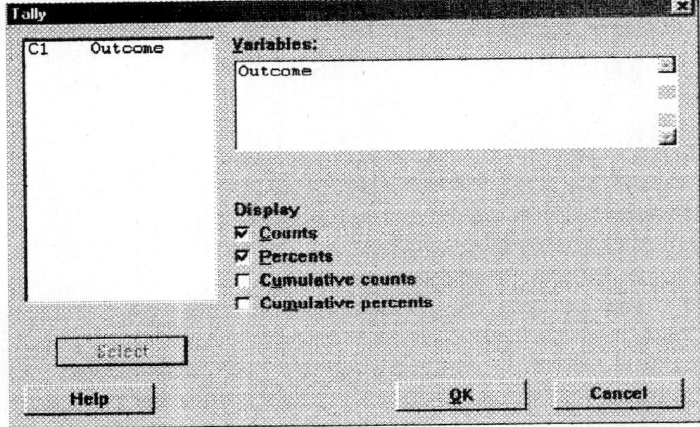

The results are

Summary Statistics for Discrete Variables

Outcome	Count	Percent
1	17	17.00
2	14	14.00
3	19	19.00
4	16	16.00
5	16	16.00
6	18	18.00

If you have a finite population, and wish to sample from it, you may use the command SAMPLE. This command requires that your population already be stored in a column.

➤Calc ➤ Random Data ➤ Sample from a Column

Dialogue Box Responses

Sample ___ rows from columns: Provide sample size and list column number C# containing population

Store sample in: Provide column number C# where you want to store the sample items.

Example

Take a sample of size 10 without replacement from the population of numbers 1 through 200.

First we need to enter the numbers 1 through 200 in column C1. The easiest way to do this is to use the patterned data option.

➤ Calc ➤ Make Patterned Data ➤ Simple Set of Numberes

Dialogue Box Resonses

Store patterned data in: List column number

From first number: 1 for this example

To last value: 200 for this example

In steps of: 1 for this example

Tell how many times to lest each value or sequence

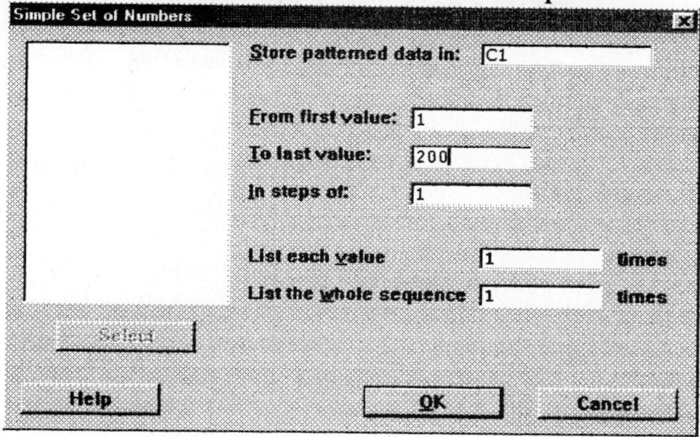

Next we use the **➤Calc ➤ Random Data ➤ Sample from a Column** choice to take a sample of 10 items from C1 and store them in C2

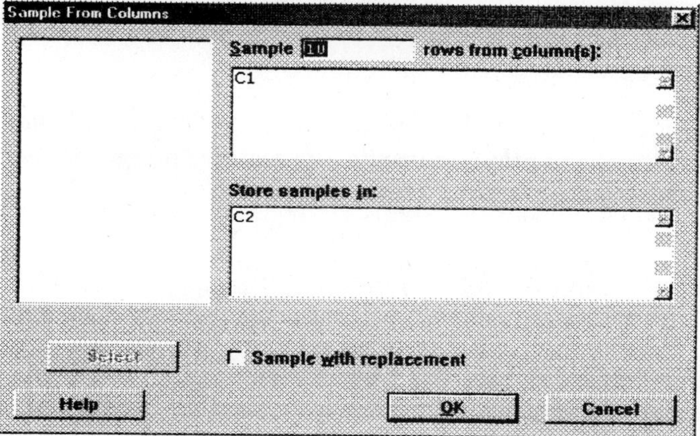

Finally, go to the Data Window and label C2 as Sample. Use ➤**Manip** ➤ **Display Data**
The results are

Data Display

```
Sample
      95     17     158     39     68     100     49     59     149     143
```

Lab Activities for Random Samples

> These activities coordinate with Section 2.1 Random Samples
> *Understandable Statistics*

1. Out of a population of 8173 eligible county residents, select a random sample of 50 for prospective jury duty. Should you sample with or without replacement? Use the RANDOM command with subcommand INTEGER A = 1 to B = 8173. Use the SORT command to sort the data so that you can check for repeated values. If necessary, use the RANDOM command again to continue sampling until you have 50 different people.

2. Retrieve the Minitab worksheet WEIGHTS.MTP on the data disk. This file contains weights of a random sample of linebackers on professional football teams. The data is in Column 1. Use the SAMPLE command to take a random sample of 10 of these weights. Print the 10 weights included in the sample.

Simulating experiments in which outcomes are equally likely is another important use of random numbers.

3. We can simulate dealing bridge hands by numbering the cards in a bridge deck from 1 to 52. Then we draw a random sample of 13 numbers without replacement from the population of 52 numbers. A bridge deck has 4 suits: hearts, diamonds, clubs, and spades. Each suit contains 13 cards; those numbered 2 through 10, a jack, a queen, a king, and an ace. Decide how to assign the numbers 1 through 52 to the cards in the deck.

   a) Use the RANDOM command with INTEGER subcommand to get the numbers of the 13 cards in one hand. Translate the numbers to specific cards and tell what cards are in the hand. For a second game, the cards would be collected and reshuffled. Use the computer to determine the hand you might get in a second game.

   b) Store the 52 cards in C1, and then use the SAMPLE command to sample 13 cards. Put the results in C2, name C2 as 'my hand' and print the results. Repeat this process to determine the hand you might get in a second game.

   c) Compare the four hands you have generated. Are they different? Would you expect this result?

4. We can also simulate the experiment of tossing a fair coin. The possible outcomes resulting from tossing a coin are heads or tails. Assign the outcome heads the number 2 and the outcome tails

the number 1. Use RANDOM with INTEGER subcommand to simulate the act of tossing a coin 10 times. Use TALLY with COUNTS and PERCENTS subcommands to tally the results. Repeat the experiment with 10 tosses. Do the percents of outcomes seem to change? Repeat the experiment again with 100 tosses.

HISTOGRAMS AND FREQUENCY DISTRIBUTIONS

MINITAB has graphics in two modes. The default mode is high resolution or professional graphics. There is also an option to use character graphics which appear in text mode. We will use the high resolution mode.

➤Graph ➤Histogram

Dialogue Box Responses

Graph variables: Column containing data

Data display: Bar for each Graph

Click on [Options] and select

Frequency for type of graph

Cutpoints for type of interval

Midoint/cutpoint positions for Defininition of intervals.

List the class boundaries (as computed in *Understandable Statistics*)

Note: If you do not use Options, the computer sets number of classes automatically. It uses the convention that data falling on a boundary are counted in the class below the boundary.

Example

Let's make a histogram of the data we stored in the worksheet ADS (created in Chapter 1). We'll use C1, the column with the number of ads per hour on prime time TV. Use four classes.

First we need to retrieve the worksheet. Use **➤File ➤ Open Worksheet**. Scroll to the drive containing the worksheet. We used 3½ disk drive [A]. Click on the file.

The number of ads per hour of TV is in column C1. Use **➤Graph ➤Histogram**. The dialogue boxes follow.

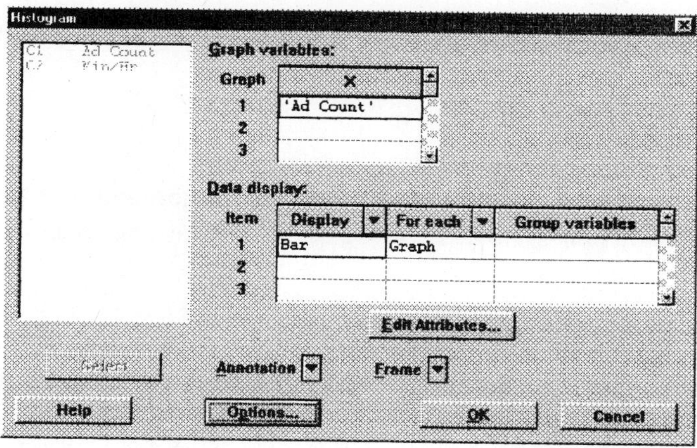

Note that the low data value i 13 and the high is 28. Using techniques shown in the text *Understandable Statistics*, we see that the class boundaries for 4 classes are 12.5; 16.5; 20.5; 24.5; 28.5. These values are listed in the Definition of Intervals. Use spaces to separate the entries.

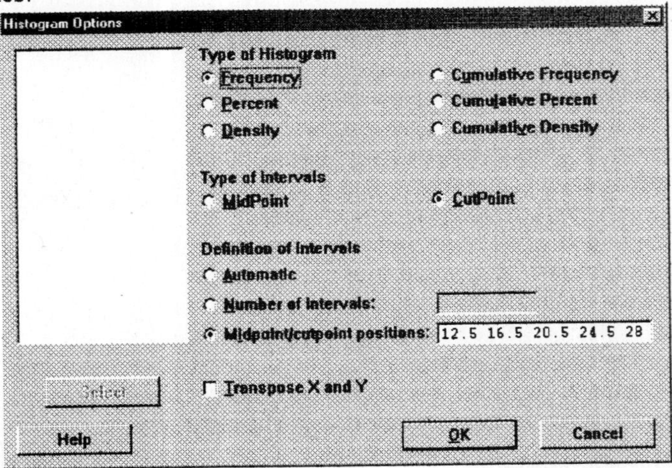

The result follows:

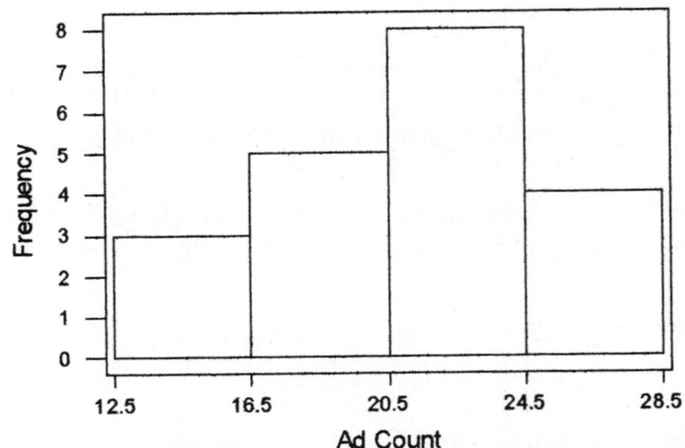

Lab Activities for Histograms

1. The ADS worksheet contains a second column of data that records the number of minutes per hour consumed by ads during prime time TV. Retrieve the ADS worksheet again and use column C2 to
   a) make a histogram letting the computer scale it.
   b) sort the data and find the smallest data value.
   c) make a histogram using the smallest data value as the starting value and an increment of 4 minutes. Do this by using cutpoints, with the smallest value as the first cutpoint and incrementing cutpoints by 4 units.

Lab Activities for Histograms continued

2. As a project for her finance class Linda gathered data about the number of cash requests made at an automatic teller machine located in the student center between the hours of 6PM and 11PM. She recorded the data every day for four weeks. The data follow.

25 17 33 47 22 32 18 21 12 26 43 25
19 27 26 29 39 12 19 27 10 15 21 20
34 24 17 18

a) Enter the data.
b) Use the command HISTOGRAM to make a histogram.
c) Use the SORT command to order the data and identify the low and high values. Use the low value as the start value and an increment of 10 to make another histogram.

3. Choose one of the following files from the data disk.

DISNEY STOCK VOLUME: **DISN.mtp**
WEIGHTS OF PRO FOOTBALL PLAYERS: **WEIGHTS.mtp**
HEIGHTS OF PRO BASKETBALL PLAYERS: **HEIGHTS.mtp.**
MILES PER GALLON GASOLINE CONSUMPTION: **MPGAL.mtp**
FASTING GLUCOSE BLOOD TESTS: **GLUCOS.mtp**
NUMBER OF CHILDREN IN RURAL CANADIAN FAMILIES: **CHILD.mtp**

a) Make a histogram letting MINITAB scale it
b) Make a histogram using five classes.

4. Histograms are not effective displays for some data. Consider the data

1 2 3 6 7 4 7 9 8 4 12 10
1 9 1 12 12 11 13 4 6 206

Enter the data and make a histogram letting MINITAB do the scaling. Next scale the histogram with starting value 1 and increment 20. Where do most of the data values fall? Now drop the high value 206 from the data. Do you get more refined information from the histogram by eliminating the high and unusual data value?

Stem-and-Leaf Displays

MINITAB supports may of the exploratory data analysis methods. You can create a stem-and-leaf display with the following menu choices.

➤Graph ➤ Stem-and-Leaf
Dialogue Box Responses
Variables: Column numbers C# containing the data
Increment: difference in value between smallest possible data in any adjacent lines.
Choose increment 10 for 1 line per stem or 5 for 2 lines per stem.

Example

Let's take the data in the worksheet ADS and make a stem-and-leaf display of C1. Recall the C1 contains the number of ads occurring in anhour of prime time TV.

Use the menu **➤Graph ➤ Stem-and-Leaf**.

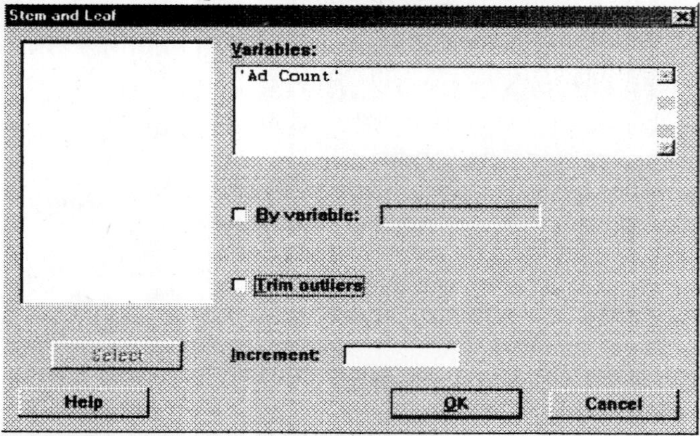

The increment defaulted to 2, so leaf units 0 1 are on one line 2 3 on the next, 4 5 on the next,etc. The results follow

Character Stem-and-Leaf Display

```
Stem-and-leaf of Ad Count   N  = 20
Leaf Unit = 1.0

    2      1 33
    3      1 5
    4      1 7
    7      1 899
    9      2 01
   (6)     2 222333
    5      2 455
    2      2 7
    1      2 8
```

The first column gives the depth of the data. That is it counts the number data cumulated in each line from the top until we reach the line containing the middle value, which is indicated in this example as (6). This indicates that the middle data value is in this line which contains 6 data values. The other numbers indicate the number of data cumulated from the bottom. The second column gives the stem and the last gives the leaves.

Let's remake a stem leaf with 2 lines per stem. That means the leaves 0 - 4 are on one line and leaves 5-9 are on the next. The difference in smallest possible leaves per adjacent lines is 5. Therefore set the increment as 5.

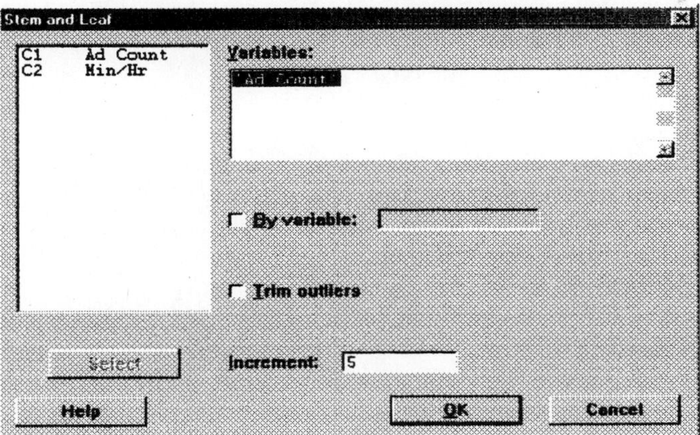

The results are

Character Stem-and-Leaf Display

```
Stem-and-leaf of Ad Count   N  = 20
Leaf Unit = 1.0

    2     1 33
    7     1 57899
   (9)    2 012223334
    4     2 5578
```

Lab Activities for Stem-and Leaf

(Section 2.4 of *Understandable Statistics)*

1. Retrieve worksheet ADS again, and make a stem-and-leaf display of the data in C2. This data gives the number of minutes of ads per hour during prime tie TV programs.
   a) Use and increment of 2
   b) Use and increment of 5

2. In a physical fitness class students ran 1 mile on the first day of class. These are their times in minutes.

12	11	14	8	8	15	12	13	12
10	8	9	11	14	7	14	12	9
13	10	9	12	12	13	10	10	9
12	11	13	10	10	9	8	15	17

a) Enter the data in a worksheet
b) Make a stem-and-leaf display and let the computer set the increment.
c) Use the TRIM option and let the computer set the increment. How does this display differ from the one in part b?
d) Set your own increment and make a stem-and-leaf display.

COMMAND SUMMARY

To generate a random sample

RANDOM K into C...C selects a random sample from the distribution described in the subcommand
WINDOWS menu: **Calc ➤ Random data**

INTEGER K to K distribution of integers from K to K

Other distributions that may be used with the RANDOM command. We will study many of these in later chapters.

BERNOULLI P = K
BINOMIAL N = K, P = K
CHISQUARE degrees of freedom = K
DISCRETE values in C probabilities in C
F df numerator = K, df denominator = K
NORMAL [mean = K [standard deviation = K]]
POISSON mean = K
T degrees of freedom = K
UNIFORM continuous distribution on [K to K]

SAMPLE K rows from C...C and put results in C...C takes a random sample of rows without replacement
REPLACE causes the sample to be taken with replacement

To organize data

SORT C...C put in C...C sorts the data in the first column and carries the other columns along
WINDOWS menu: **Manip ➤ Sort**

DESCENDING C...C subcommand to sort in descending order

TALLY data in C...C tallies data in columns with integers
COUNTS
PERCENTS
CUMCOUNTS
CUMPERCENTS
ALL gives all four values
WINDOWS menu: **Stats ➤ Tables ➤ Tally**

HISTOGRAM C...C prints a separate histogram for data in each of the listed columns

START with midpoint = K [end with midpoint = K]

INCREMENT = K specifies distance between midpoints

WINDOWS menu: (for professional graphics) **Graph ➤ Histogram (options for cutpoints)**

WINDOWS menu: (for character graphics) **Graph ➤ Character Graphs ➤ Histogram**

STEM-AND-LEAF display of C...C makes separate stem-and-leaf displays of data in each of the listed columns

INCREMENT = K sets the distance between two display lines

TRIM outliers lists extreme data on special lines

WINDOWS menu: **Graph ➤a Chracter Graphs ➤ Stem-and-Leaf**

CHAPTER 3 AVERAGES AND VARIATION

AVERAGES AND STANDARD DEVIATION OF UNGROUPED DATA

Sections 3.1 and 3.2 of UNDERSTANDABLE STATISTICS describe some of the basic summary statistics that are useful. The command DESCRIBE of MINITAB gives many of these values.

➤**Stat** ➤ **Basic Statistics** ➤ **Display Descriptive Statistics** prints descriptive statistics for each column of data.

Dialogue Box Response

Variables: List the columns C1 .... CN that contain the data

Graphs option: You may print histograms, etc directly from this menu

The labels are for Display Descriptive Statistics are

N number of data in C

N* number of missing data in C

MEAN arithmetic mean of C

MEDIAN median or center of the data in C

TRMEAN 5% trimmed mean (the mean obtained after removing the smallest 5% and largest 5% of the data)

STDEVthe sample standard deviation of C, s

SEMEAN standard error of the mean, STDEV/SQRT(N) (we will use this value in Chapter 7)

MIN minimum data value in C

MAX maximum data value in C

Q1 lst quartile of distribution in C

Q3 3rd quartile of distribution in C (Q1 and Q3 are similar to Q_1 and Q_3 as discussed in Section 3.4 of UNDERSTANDABLE STATISTICS. However, the computation process is slightly different and give values slightly different from those in the text.)

Example

Let's again consider the data about the number and duration of ads during prime time TV. We will retrieve worksheet ADS and use DESCRIBE on C2, the number of minutes per hour of ads during prime time TV.

First use ➤**File** ➤**Open Worksheet** to open worksheet ADS

Next use **➤Stat ➤ Basic Statistics ➤ Display Descriptive Statistics** and click on **OK**.

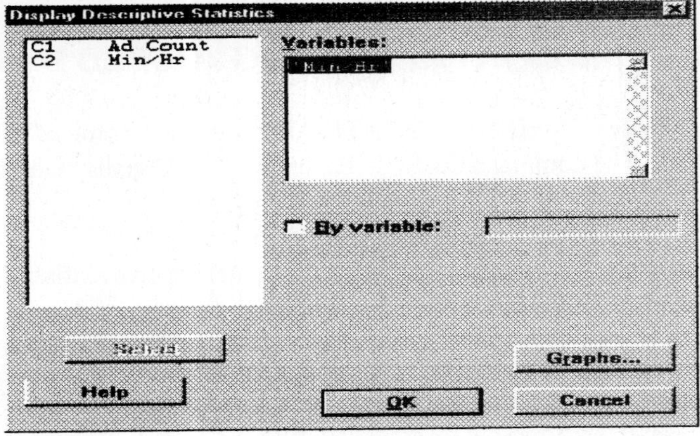

The results follow.

Descriptive Statistics

```
Variable          N      Mean    Median    TrMean     StDev   SE Mean
Min/Hr           20    11.675    11.100    11.583     1.849     0.413

Variable      Minimum   Maximum        Q1        Q3
Min/Hr          9.300    15.700    10.350    12.225
```

ARITHMETIC IN MINITAB

The standard deviation given in STDEV is the sample standard deviation

$$s=\sqrt{\frac{\Sigma(x - \bar{x})^2}{N - 1}}$$

We can compute the population standard deviation σ by multiplying s by the factor shown below

$$\sigma=s\sqrt{\frac{N - 1}{N}}$$

MINITAB allows us to do such arithmetic. Use the built in calculator under menu selection **➤Calc ➤ Calculator**. Note that * means multiply and ** means exponent.

Example

Let's use the arithmetic operations to evaluate the population standard deviation and population variance for the minutes per hour of TV ads. Notice that the sample standard deviation s = 1.849 and the sample size is 20.

Use the CALCULATOR as follows: Select menus ➤**Calc** ➤ **Calculator**
Then enter the expression for the population variance on the calculator

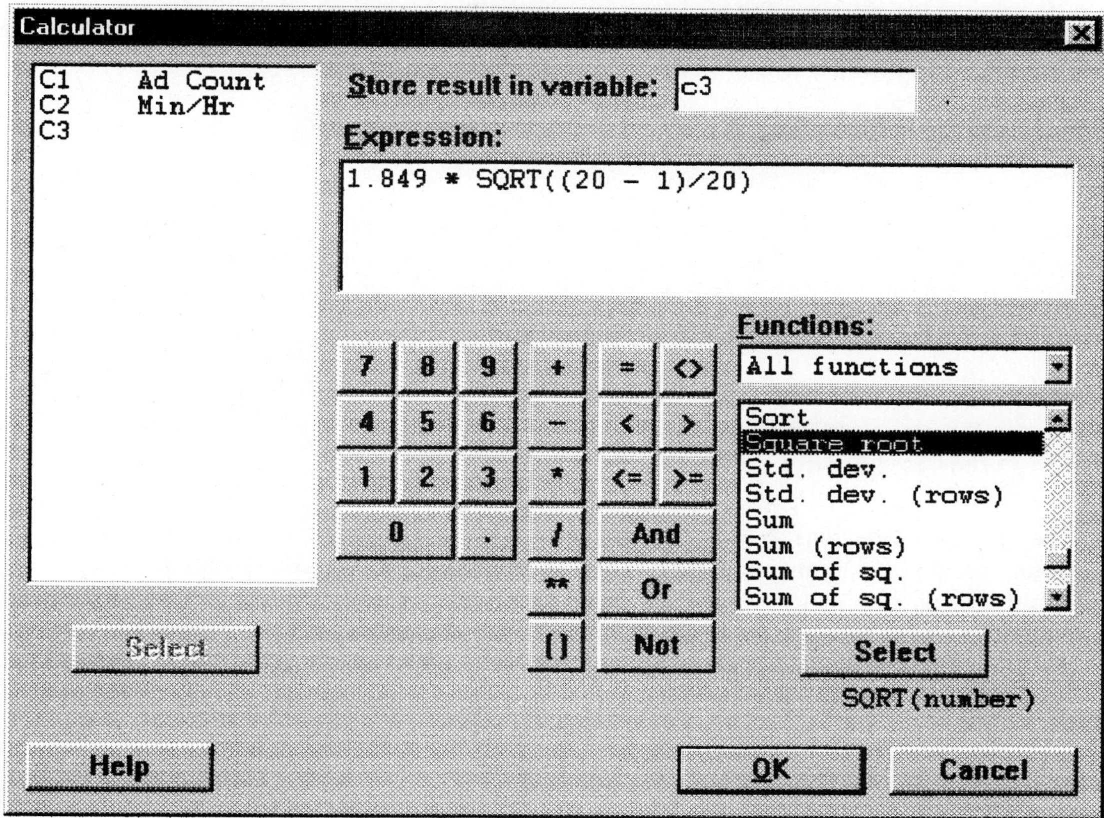

Note that we access the square root function in the function menu. The result is stored in C3. Go to the Data Window and name C3 as PopStDv. Then use ➤**Manip** ➤**Display Data**. The result follows

Data Display

```
PopStDv
 1.80218.
```

As shown in Chapter 1, you can also use the MINITAB calculator to do arithmetic with columns of data. For instance, to add the number 5 to each entry in column C2 and store results in C3, simply do the following in the Expression box

type C2   click on + key   type 5

Designate C3 as the column in which to store the results.

Note that you can store a single number as a constant designated K# instead in a column.

Lab Activities for Averages and Variation

(Sections 3.1 and 3.2 of *Understandable Statistics*)

1. A random sample of 20 people were asked to dial 30 telephone numbers each. The incidence of numbers misdialed by these people follow:

   3 2 0 0 1 5 7 8 2 6
   0 1 2 7 2 5 1 4 5 3

   Enter the data and use the menu selections ➤ **Basic Statistics** ➤ **Display Descriptive Statistics** to find the mean, median, minimum value, maximum value, and standard deviation.

2. Consider the test scores of 30 students in a political science class.

   85 73 43 86 73 59 73 84 100 62
   75 87 70 84 97 62 76 89 90 83
   70 65 77 90 94 80 68 91 67 79

   a) Use the menu selections ➤ **Basic Statistics** ➤ **Display Descriptive Statistics** to find the mean, median, minimum value, maximum value, and sample standard deviation.

   b) Greg was in a political science class and suppose he missed a number of classes because of illness. Suppose he took the exam anyway and made a score of 30 instead of 85 as listed in the data set. Change the 85 (first entry in the data set) to 30 and use the **DESCRIBE** command again. Compare the new mean, median and standard deviation with the ones in part (a). Which average was most affected: median or mean? What about the standard deviation?

3. Consider the 10 data values

   4 7 3 15 9 12 10 2 9 10

   a) Use the the menue selections ➤ **Basic Statistics** ➤ **Display Descriptive Statistics** to find the sample standard deviation of these data values. Then, following example 2 as a model, find the population standard deviation of these data. Compare the two values.

   b) Now consider these 50 data values in the same range.

   7 9 10 6 11 15 17 9 8 2
   2 8 11 15 14 12 13 7 6 9
   3 9 8 17 8 12 14 4 3 9
   2 15 7 8 7 13 15 2 5 6
   2 14 9 7 3 15 12 10 9 10

   Again use the menu selections ➤ **Basic Statistics** ➤ **Display Descriptive Statistics** to find the sample standard deviation of these data values. Then, as above, find the population standard deviation of these data. Compare the two values.

   c) Compare the results of parts (a) and (b). As the sample size increases, does it appear that the difference between the population and sample standard deviations decreases? Why would you expect this result from the formulas?

4. In this problem we will explore the effects of changing data values by multiplying each data value by a constant, or by adding the same constant to each data value.

a) Make sure you have a new Then enter the following data into C1

1 8 3 5 7 2 10 9 4 6 3

Use the menu selections ➤ **Basic Statistics** ➤ **Display Descriptive Statistics** to find the mean, median, minimum and maximum values and sample standard deviation.

b) Now use the calculator box to create a new column of data C2 = 10*C1. Use menu selections again to find the mean, median, minimum and maximum values, and sample standard deviation of C2. Compare these results to those of C1. How do the means compare? How do the medians compare? How do the standard deviations compare? Referring to the formulas for these measures (see Sections 3.1 and 3.2 of UNDERSTANDABLE STATISTICS) can you explain why these statistics behaved the way they did? Will these results generalize to the situation of multiplying each data entry by 12 instead of by 10? Confirm your answer by creating a new C3 that has each datum of C1 multiplied by 12. Predict the corresponding statistics that would occur if we multiplied each datum of C1 by 1000. Again, create a new column C4 that does this and use DESCRIBE to confirm your prediction.

c) Now suppose we add 30 to each data value in C1. We can do this by using the calculator box to create a new column of data C6 = C1 + 30. Use menu selection ➤ **Basic Statistics** ➤ **Display Descriptive Statistics** on C6 and compare the mean, median, and standard deviation to those shown for C1. Which are the same? Which are different? Of those that are different, did each change by being 30 more than the corresponding value of part (a)? Again look at the formula for standard deviation. Can you predict the observed behavior from the formulas? Can you generalize these results? What if we added 50 to each datum of C1? Predict the values for the mean, median, and sample standard deviation. Confirm your predictions by creating a column C7 in which each datum is 50 more than that in the respective position of C1. Use menu selections ➤ **Basic Statistics** ➤ **Display Descriptive Statistics** on C7.

d) Name C1 as 'orig', C2 as 'T10', C3 as 'T12', C4 as 'T1000',C6 as 'P30', C7 as 'P50'. Now use the menu selections ➤ **Basic Statistics** ➤ **Display Descriptive Statistics** C1-C4 C6 C7 and look at the display. Is it easier to compare the results this way?

BOX-AND-WHISKER PLOTS

The box-and-whisker plot is another of the explanatory data analysis techniques supported by MINITAB. These plots are discussed in Section 3.4 of UNDERSTANDABLE STATISTICS.

MINITAB uses hinges rather than quartiles Q_1 and Q_3. The lower hinge HL is close in value to Q_1 and the upper hinge HU is close in value to Q_3.

To identify unusual observations, fences located as follows are used.

inner fence HL - 1.5(HU - HL) and HU + 1.5(HU - HL)
outer fence HL - 3.0(HU - HL) AND HU + 3.0(HU - HL)

Whiskers extend to the most extreme observation within the inner fence. Values between the inner and outer fence are designated by a * while those beyond the outer fences are designated by a 0.

The menu selections are
➤ **Graph** ➤ **Boxplot**
Dialogue Box Responses: Under Y, enter the column number C# containing the data
Annotation: Open box and you can title the graph
Display: IQ Range Box with Outliers shown
There are other boxes available within this box. See Help to learn more about these options

Example

Now let's make a box-and-whisker plot of the data stored in worksheet ADS. C1 contains the number of ads per hour of prime time TV while C2 contains the duration per hour of the ads.

Use the menu selection ➤ **Graph** ➤ **Boxplot**
C1 for Y. Leave X blank.. Click on OK

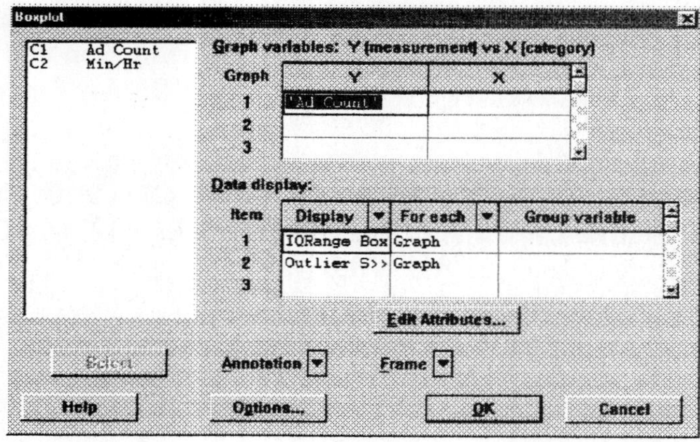

The results follow

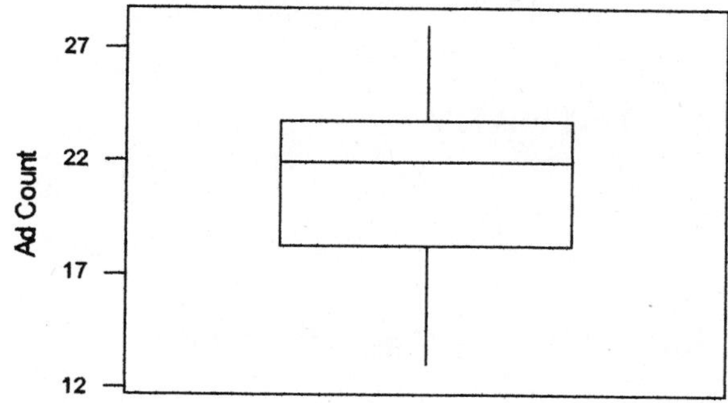

Lab Activities for Box-and-Whisker Plots

(Section 3.4 of *Understandable Statistics)*

1. State regulated nursing homes have a requirement that there be a minimum of 132 minutes of nursing care per resident per 8 hr shift. During an audit of Easy Life Nursing home, a random sample of 30 shifts showed the number of minutes of nursing care per resident per shift to be

200	150	190	150	175	90	195	115	170	100
140	270	150	195	110	145	80	130	125	115
90	135	140	125	120	130	170	125	135	110

a) Enter the data.
b) Make a box-and-whisker plot. Are there any unusual observations?
c) Make a stem-and-leaf plot. Compare the two ways of presenting the data.
d) Make a histogram. Compare the information in the histogram with that in the other two displays.
e) Use the ➤**Stat** ➤**Basic Statistics** ➤ **Display Descriptive Statistics** menu selections.
f) Now remove any data beyond the outer fences. Do this by inserting an asterisk * in place of the number in the data cell. Use the menu selections ➤**Stat** ➤**Basic Statistics** ➤ **Display Descriptive Statistics** on this data. How do the means compare?
h) Pretend that you are writing a brief article for a newspaper. Describe the information about the time nurses spend with residents of a nursing home. Use non technical terms. Be sure to make some comments about the "average" of the data measurements and some comments about the spread of the data.

2. Select one of these data files from the DATA DISK and repeat parts (b) through (h).

DISNEY STOCK VOLUME: **DISN.mtp**
WEIGHTS OF PRO FOOTBALL PLAYERS: **WEIGHTS.mtp**
HEIGHTS OF PRO BASKETBALL PLAYERS: **HEIGHTS.mtp.**

MILES PER GALLON GASOLINE CONSUMPTION: **MPGAL.mtp**
FASTING GLUCOSE BLOOD TESTS: **GLUCOS.mtp**
NUMBER OF CHILDREN IN RURAL CANADIAN FAMILIES: **CHILD.mtp**

COMMAND SUMMARY

To summarize data by column

DESCRIBE C...C prints descriptive statistics
WINDOWS menu:➤**Stat** ➤**Basic Statistics** ➤ **Display Descriptive Statistics**

COUNT **C [put into K]** counts the values
N **C [put into K]** counts the non-missing values
NMIS **C [put into K]** counts the missing values
SUM **C [put into K]** sums the values
MEAN **C [put into K]** gives arithmetic mean of values
STDEV **C [put into K]** gives sample standard deviation
MEDIAN **C [put into K]** gives the median of the values
MINIMUM **C [put into K]** gives the minimum of the values
MAXIMUM **C [put into K]** gives the maximum of the values
SSQ **C [put into K]** gives the sum of squares of values

To summarize data by row

RCOUNT E...E put into C
RN E...E put into C
RNMIS E...E put into C
RSUM E...E put into C
RMEAN E...E put into C
RSTDEV E...E put into C
RMEDIAN E...E put into C
RMIN E...E put into C
RMAX E...E put into C
RSSQ E...E put into C

To display data

BOXPLOT C makes a box-and-whisker plot of data in column C
START = K [end = k]
INCREMENT = K
WINDOWS menu: (professional graphics) **Graph** ➤ **Boxplot**

To do arithmetic

LET E = expression evaluates the expression and stores the result in E where E may be a column or a constant

- ** raises to a power
- * multiplication
- / division
- + addition
- - subtraction

SQRT(E) takes the square root
ROUND(E) rounds numbers to the nearest integer
There are other arithmetic operations possible.

WINDOWS menu selections: **➤Calc ➤ Calculator**

CHAPTER 4 ELEMENTARY PROBABILITY

RANDOM VARIABLES AND PROBABILITY DISTRIBUTIONS

MINITAB supports random samples from a column of numbers or from many probability distributions. See the options under **➤Calc ➤Random Data**. By using some of the same techniques shown in Chapter 2 of this guide for random samples, you can simulate a number of probability experiments.

Example

Simulate the experiment of tossing a fair coin 200 times. Look at the percent of heads and the percent of tails. How do these compare with the expected 50% for each.

Assign the outcome heads to digit 1 and tails to digit 2. We will draw a random sample of size 200 from the distributions of integers from a minimum of 1 to a maximum of 2.

Use the menu selections **➤Calc ➤ Random Data ➤ Integer**. In the dialogue box, enter 200 for the number of rows and 1 for the minimum and 2 for the maximum. Put the data in column C1 and label the column Coin.

To tally the results use **➤Stat ➤Table ➤ Tally** and check the counts and percents options.

The results are

Summary Statistics for Discrete Variables

```
Coin  Count  Percent
   1    101    50.50
   2     99    49.50
  N=    200
```

Lab Activities in Probability

1. Use the RANDOM command and INTEGER A = 0 to B = 1 subcommand to simulate 50 tosses of a fair coin. Use the TALLY command with COUNT and PERCENT subcommands to record the percent of each outcome. Compare the result with the theoretical expected percents (50% heads, 50% tails). Repeat the process for 1000 trials. Are these outcomes closer the results predicted by the theory?

2. We can use the RANDOM 50 C1 C2 command with INTEGER A = 1 to B = 6 subcommand to simulate the experiment of rolling two dice 50 times and recording each sum. This command puts outcomes of die 1 into C1 and those of die 2 into C2. Put the sum of the die into C3. Then use the TALLY command with COUNT and PERCENT subcommands to record the percent of each outcome. Repeat the process for 1000 rolls of the dice.

CHAPTER 5 THE BINOMIAL DISTRIBUTION AND RELATED TOPICS

THE BINOMIAL PROBABILITY DISTRIBUTION

The binomial probability distribution is discussed in Chapter 5 of *Understandable Statistics*. It is a discrete probability distribution controlled by the number of trials n and the probability of success on a single trial, p.

MINITAB has three main commands for studying probability distributions.

The PDF (probability density function) gives the probability of a specific value for a discrete distribution.

The CDF (cumulative distribution function) for a value X gives the probability a random variable with the distribution specified in a subcommand is less than or equal to X.

The INVCDF gives the inverse of the CDF. In other words, for a probability P, INVCDF returns the value X such that P ≈ CDF(X). In the case of a binomial distribution, INVCDF often gives the two values of X for which P lies between the respective CDF(X).

The three commands PDF, CDF, and INVCDF apply to many probability distributions. To apply them to a binomial distribution, we need to use the menu selections

➤Calc ➤ Probability distribution ➤Binomial

Dialogue Box Responses

Select Probability for PDF; Cumulative probability for CDF; Inverse cumumlative probability for INVCDF

Number of trials: use the value of n in a binomial experiment

Probability of success: use the value of p, the probability of success on a single trial

Input column: put the values for r, the number of successes in a binomial experiment in a column such as C1. Select an optional storage column.

Note: MINITAB uses X instead of r to count the number of successes

Input constant: Instead of putting values of r in a column, you can type a specific value of r in the dialogue box.

Example

A surgeon performs a difficult spinal column operation. The probability of success of the operation is p = 0.73. Ten such operations are scheduled. Find the probability of success for 0 through 10 successes out of these operations.

First enter the possible values of r, 0 through 10 in C1 and name the column as r. We will put the probabilities in C2, so name the column P(r).

Fill in the Dialogue Box as shown on the next page

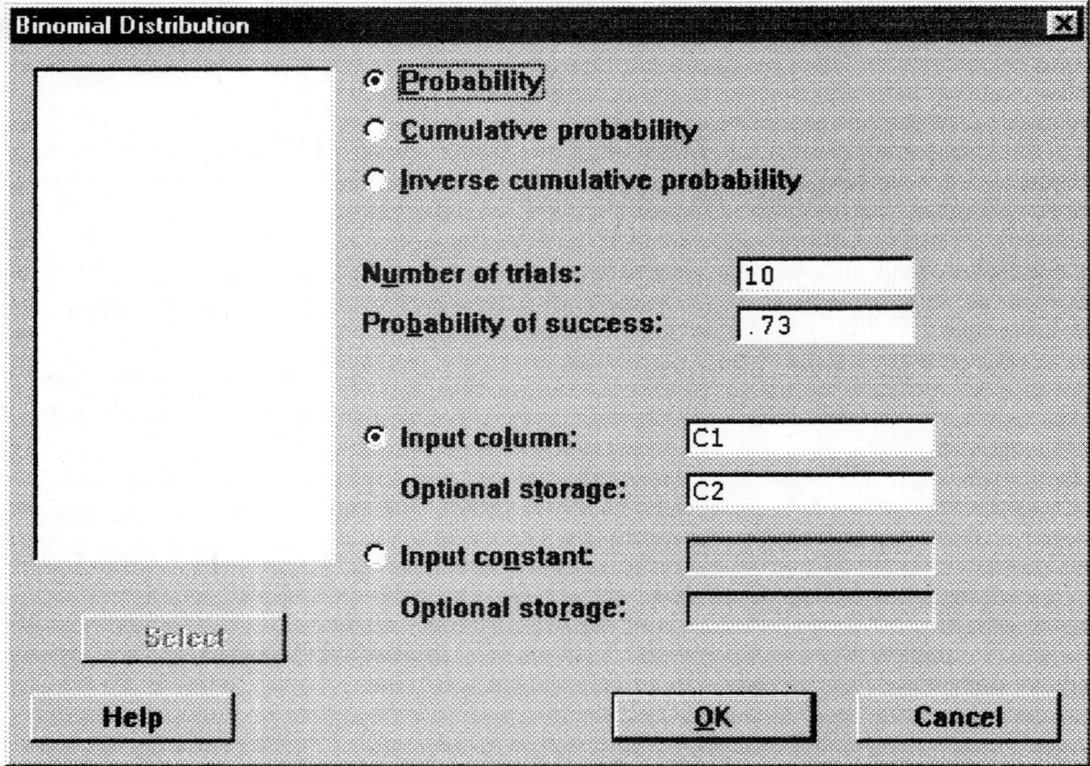

Then use the ➤**Manip** ➤**Display data** command.

Data Display

Row	r	P(r)
1	0	0.000002
2	1	0.000056
3	2	0.000677
4	3	0.004883
5	4	0.023104
6	5	0.074961
7	6	0.168893
8	7	0.260935
9	8	0.264559
10	9	0.158953

Next use the CDF command to find the probability of 5 or fewer successes. In this case use the option for an input constant of 5. The output will be $P(r \leq 5)$. Note that MINITAB uses X in place of r.

The results follow:

Cumulative Distribution Function

```
Binomial with n = 10 and p = 0.730000

       x         P( X <= x)
    5.00            0.1037
```

Finally use the INVCDF to determine how many operations should be performed in order for the probability of that many or fewer successes to be 0.5. We select Inverse. Use .5 as the input constant.

The results follow

Inverse Cumulative Distribution Function

```
Binomial with n = 10 and p = 0.730000

        x       P( X <= x)                    x          P( X <= x)
        6          0.2726                     7             0.5335
```

Lab Activities for Binomial Distributions

1. You toss a coin 8 times. Call heads success. If the coin is fair, the probability of success P is 0.5. What is the probability of getting exactly 5 heads out of 8 tosses? of exactly 20 heads out of 100 tosses?

2. A bank examiner's record shows that the probability of an error in a statement for a checking account at Trust Us Bank is 0.03. The bank statements are sent monthly. What is the probability that exactly two of the next 12 monthly statements for our account will be in error? Now use the CDF option to find the probability that <u>at least</u> two of the next 12 statements contain errors. Use this result with subtraction to find the probability that <u>more than</u> two of the next 12 statements contain errors. You can use the Calculator key to do the required subtraction.

3. Some tables for the binomial distribution give values only up to 0.5 for the probability of success p. There is a symmetry to the values for p greater than 0.5 with those values of p less than 0.5.

   a) Consider the binomial distribution with n = 10 and p = .75. Since there are 0-10 successes possible, put 0 - 10 in C1. Use PDF option with C1 and store the distribution probabilities in C2. Name C2 = 'P=.75'. We will print the results in part (c).

   b) Now consider the binomial distribution with n = 10 and p = .25. Use PDF option with C1 and store the distribution probabilities in C3. Name C3 = 'P=.25

c) Now display C1 C2 C3 and see if you can discover the symmetries of C2 with C3. How does P(K = 4 successes with p = .75) compare to P(K = 6 successes with p = .25)?

d) Now consider a binomial distribution with n = 20 and p = 0.35. Use **PDF** on the number 5 to get P(K = 5 successes out of 20 trials with p = .35). Predict how this result will compare to the probability P(K = 15 successes out of 20 trials with p = .65). Check your prediction by using the **PDF** on 15 opiton with the binomial distribution n = 20 p = .65.

The INVCDF command for a binomial distribution can be used in the solution of Quota problems as described in Section 5.3 of *Understandable Statistics.*

4. Consider a binomial distribution with n = 25 and p = 0.64. Use the INVCDF to find the smallest number of successes K for which $P(X \leq K) = 0.98$. What is the smallest number of successes K for which $P(X \leq K) = 0.90$?

COMMAND SUMMARY

To find probabilities

PDF for values in E [put into E] calculates probabilities for the specified values of a discrete distribution and calculates the probability density function for a continuous distribution.

CDF for values in E...E [put into E...E] gives the cumulativedistribution. For any value X CDF X gives the probability that a random variable with the specified distribution has a value less than or equal to X

INVCDF for values in E [put into E] gives the inverse of the CDF.

Each of these commands apply the following distributions (as well as some others). If no subcommand is used, the default distribution is the standard normal.

BINOMIAL n = K p = K
POISSON μ = K (note that for the Poisson distribution μ = λ)
INTEGER a = K b = K
DISCRETE values in C, probabilities in C
NORMAL μ = K σ = K
UNIFORM a = K b = K
T d.f. = K
F d.f. numerator = K d.f. denominator = K
CHISQUARE d.f. = K

WINDOWS menu selection: **Calc ➤ Probability Distribution ➤ Select distribution**
In the dialogue box, select **Probability for PDF; Cumulative probability for CDF; Inverse cumulative for INV;** Enter the required information such as **E, n, p, or μ, d.f.** and so forth.

CHAPTER 6 NORMAL DISTRIBUTIONS

GRAPHS OF NORMAL DISTRIBUTIONS

The normal distribution is a continuous probability distribution determined by the value of μ and σ. The graphs of normal distributions are discussed in Section 6.1 of *Understandable Statistics.* We can sketch the graph of a normal distribution by using the menu selection

➤Calc ➤ Probability Distribution ➤Normal

Dialogue Box Resonses

Select Probability density for PDF, Cumulative probability for CDF, or Inverse cumulative probability for INVCDF

Enter the mean

Enter the standard deviation

Select an input column: Put the value of x for which you want to compute P(x) in the designated column. Designate an optional storage colun

Select an input constant: If you wish to compute P(x) for a single value x, enter that value as the constant

We will use the normal probability density option PDF to create a graph of a normal distribution with a specified mean and standard deviation.

Menu options for graphs

To graph functions in MINITAB, we use the menu

➤ Graph ➤ Plot

Dialogue Box Responses

Specify which column contains data for the X value

Specify which column contains data for the Y value

Under Display, arrow to the connect option

There are other options available. See the Help menu for more information.

Example

Graph the normal distribution with mean $\mu = 10$ and standard deviation $\sigma = 2$.

Since most of the normal curve occurs over the values $\mu - 3\sigma$ to $\mu + 3\sigma$, we will start the graph at 10 - 3(2) = 4 and end it at 0 + 3(2) = 16. We will let MINITAB set the scale on the vertical axis.

To graph a normal distribution put X value into C1 an Y values (height of the gaph at point X) into C2.

Use the menu option **➤Calc ➤Set Patterned Data ➤Simple set of numbers**. Store values in C1. First value is 4, last value is 16, increment is 0.25. Variables and sequences occur one time each. Name C1 as X.

Use **➤Calc ➤Probability Distribution ➤Normal** to generate the Y values which we will store in C2. Name C2 as P(X).

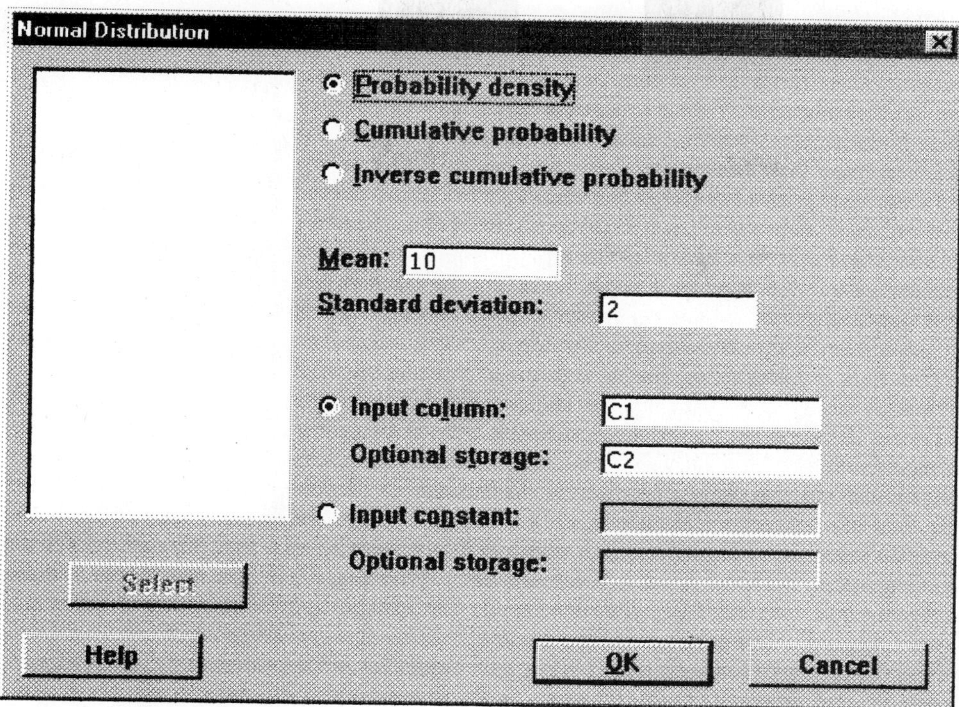

Finally, use ➤Graph ➤Plot. Set the options as shown.

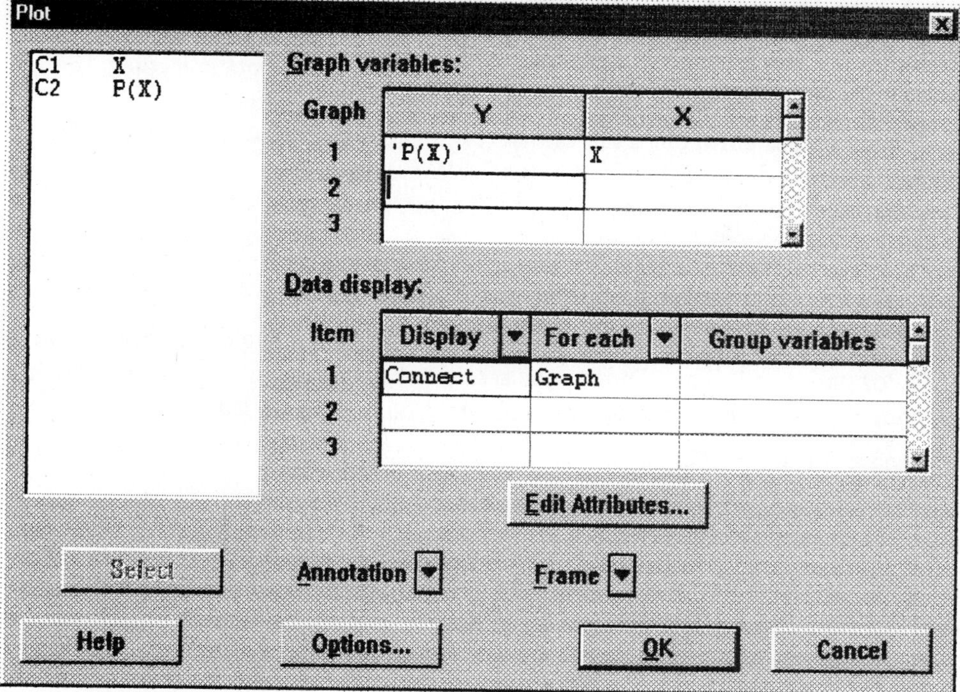

The graph follows.

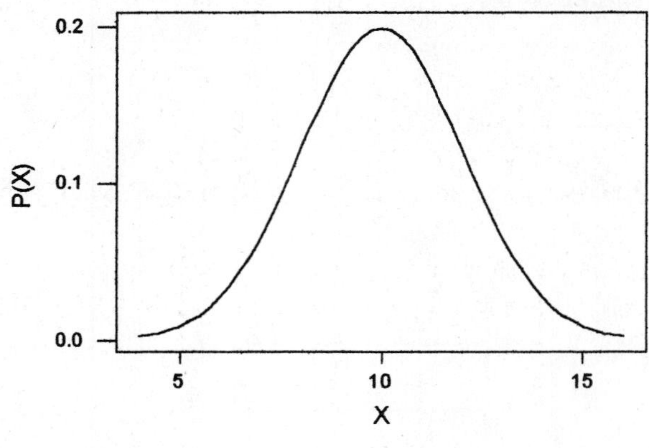

CONTROL CHARTS

MINITAB supports a variety of control charts. The type discussed in Section 6.1 of *Understandable Statistics* is called an individual chart. The menu selection is

Stat ➤ Control Chart ➤ Individual

Dialogue Box Responses

Variable: Designate the column number C1 where the data is located

Enter values for the historical mean and standard deviation

Tests option button lists out of control tests. Select numbers 1,2 and 5 for signals discussed in *Understandable Statistics*

For information about the other options, see the Help menu.

Example

In a packaging process, the weight of popcorn that is to go in a bag has a normal distribution with μ = 20.7oz and σ = 0.7 oz. During one session of packaging eleven samples were taken. Use an individual control chart to show these observations. The weights were (in oz)

19.5	20.3	20.7	18.9	9.5	20.5
20.7	21.4	21.9	22.7	23.8	

Enter the data in column C1, name the column oz.

Select **Stat ➤ Control Chart ➤ Individual**. Fill in the values for μ and σ .

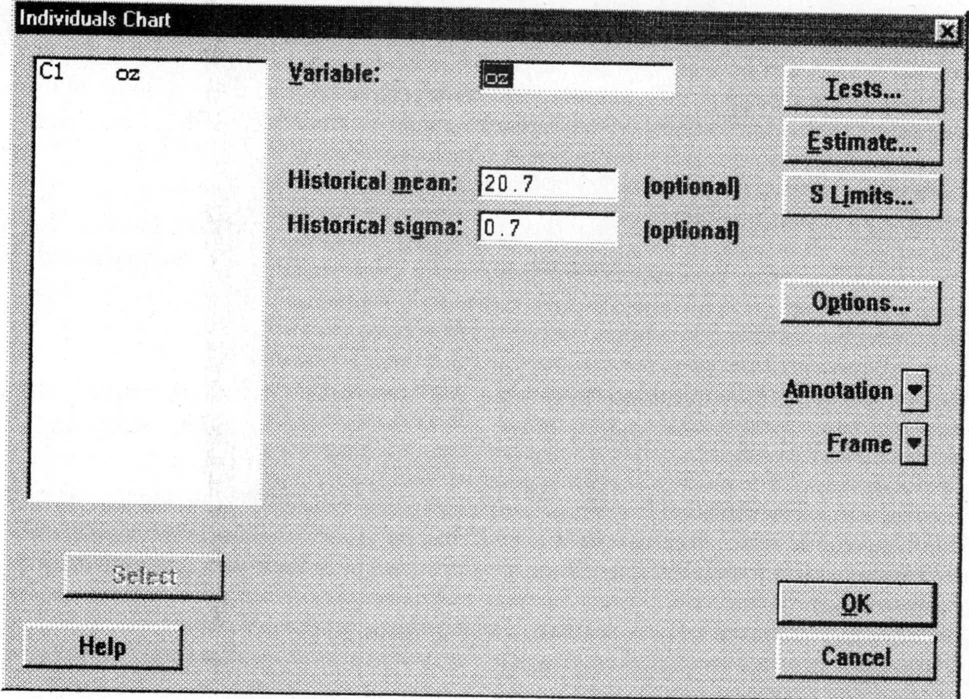

The control chart is

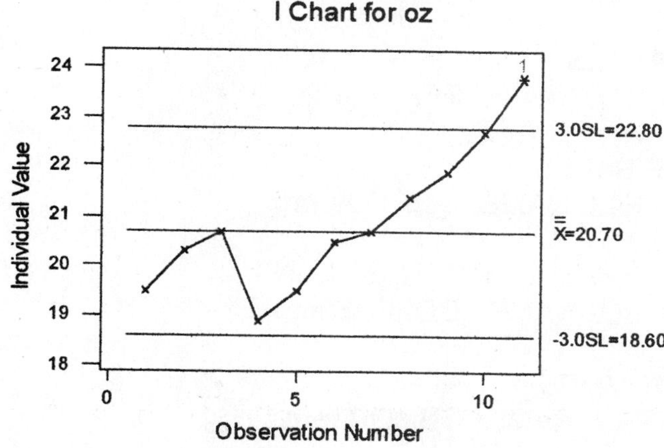

Lab Activities for Graphing Normal Distributions and Control Charts

1. a) Sketch a graph of the standard normal distribution with $\mu = 0$ and $\sigma = 1$. Generate C1 using data from -3 to 3 in increments of 0.5 (generates data from -3 to 3 with .5 increment).

   b) Sketch a graph of a normal distribution with $\mu = 10$ and $s = 1$. Generate C1 using data from -7 to 7 in increments of 0..5. Compare this graph to that of part a. Do the height and spread of the graphs appear to be the same? What is different? Why would you expect this difference.

   c) Sketch a graph of a normal distribution with $\mu = 0$ and $\sigma = 2$. Generate C1 using data from -6 to 6 in increments of 0.5. Compare this graph to that of part a. Do the height and spread of the graphs appear to be the same? What is different? Why would you expect this difference? Note, to really compare the graphs, it is best to graph them using the same scales. Redo the graph of part (a) using X from -6 to = 6. Then redo the graph in this part using the same X values as in part a and Y values ranging from 0 to high value of part (a).

2. Use one of the following MINITAB portable worksheets found on the DATA DISK. In each of the files the target value for the mean μ is stored in the C2(1) position and the target value for the standard deviation is stored in the C3(1) position.

Select from the MINITAB DATA DISK

YIELD OF WHEAT AT ROTHAMSTED EXPERIMENT STATION, ENGLAND
DATA DISK FOR MINITAB FILE NAME: WHEAT.mtp

PepsiCo STOCK CLOSING PRICES
DATA DISK FOR MINITAB FILE NAME: PEPCL.mtp

PepsiCo STOCK VOLUME OF SALES
DATA DISK FOR MINITAB FILE NAME: PEPVOL.mtp

FUTURES QUOTES FOR THE PRICE OF COFFEE BEANS
DATA DISK FOR MINITAB FILE NAME: COFFEE.mtp

INCIDENCE OF MELANOMA TUMORS
DATA DISK FOR MINITAB FILE NAME: TUMOR.mtp

PERCENT CHANGE IN CONSUMER PRICE INDEX
DATA DISK FOR MINITAB FILE NAME: CPI.mtp

Use the targeted MU and SIGMA values.

COMMAND SUMMARY

Graphing commands

Character Graphics Commands

PLOT C versus C prints a scatter plot with the first column on the vertical axis and the second on the horizontal axis. The following subcommands can be used with PLOT

TITLE = 'text' gives a title above the graph
FOOTNOTE = 'text' places a line of text below the graph
XLABEL = 'text' labels the x-axis
YLABEL = 'text' labels the y-axis
SYMBOL = 'symbol' selects the symbol for the points on the graph. The default is *
XINCREMENT = K is distance between tick marks on x-axis
XSTART = K [end = k] specifies the first tick mark and optionally the last one
YINCREMENT = K is distance between tick marks on y-axis
YSTART = K [end = K] specifies the first tick mark and optionally the last one

WINDOWS menu selection: **Graph ➤ Character Graphs ➤ Scatter Plot**
Titles, labels, and footnotes are in the **Annotate...** option
Increment and start are in the **Scale** option.

Professional Graphics

Plot C * C prints a scatter plot with the first column on the vertical axis and the second on the horizontal axis. Note that the columns must be separated by an asterisk *.

Connect connects the points with a line

Other subcommands may be used to title the graph and let the tick marks on the axes. See your MINITAB software manual for details.

WINDOWS menu selection: **Graph ➤ Plot**
Use the dialogue boxes to title the graph, label the axes, set the tick marks, and so forth. See your MINITAB software manual for details.

Control Charts

CHART C...C produces a control chart under the assumption that the data come from a normal distribution with mean and standard deviation specified by the subcommands

MU K gives the mean of the normal distribution
SIGMA K gives the standard deviation.

WINDOWS menu selection: **Stat ➤ Control Chart ➤ Individual**
Enter choices for MU and SIGMA in the dialogue box

CHAPTER 7 SAMPLING DISTRIBUTIONS

Note: This section uses session window commands instead of menu choices
CENTRAL LIMIT THEOREM

Section 7.2 of *Understandable Statistics* introduces the Central Limit Theorem. The Central Limit Theorem says that if x is a random variable with <u>any</u> distribution having mean μ and standard deviation σ then the distribution of sample means $\bar{x}$ based on random samples of size n is such that for sufficiently large n:

a) The mean of the $\bar{x}$ distribution is approximately the same as the mean of the x distribution.

b) The standard deviation of the $\bar{x}$ distribution is approximately $\sigma/\sqrt{n}$

c) The $\bar{x}$ distribution is approximately a normal distribution

Furthermore, as the sample size n becomes larger and larger, the approximations mentions in (a), (b) and (c) become better.

We can use MINITAB to demonstrate the Central Limit Theorem. The computer does not prove the theorem. A proof of the Central Limit Theorem requires advanced mathematics and is beyond the scope of an introductory course. However, we can use the computer to gain a better understanding of the theorem.

To demonstrate the Central Limit Theorem, we need a specific x distribution. One of the simplest is the <u>uniform probability distribution</u>. Let us compare the uniform distribution with the normal distribution.

Uniform Distribution and a Normal Distribution

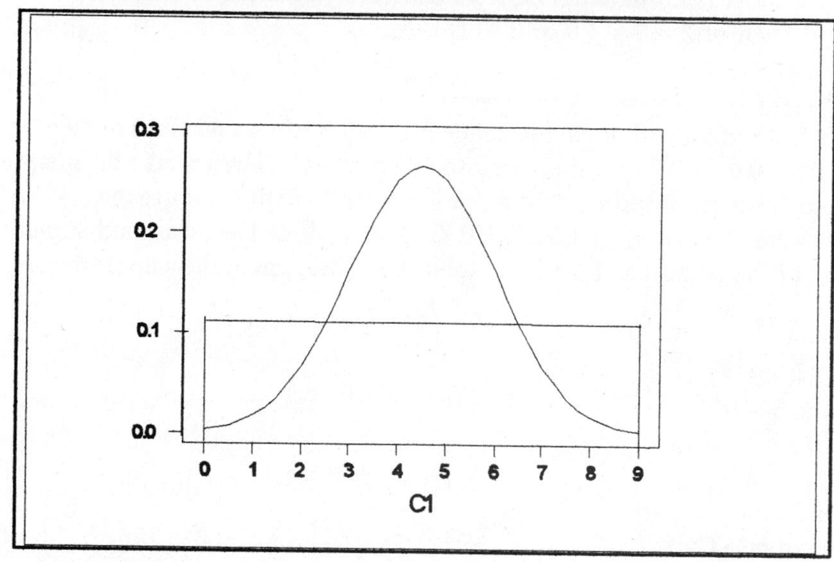

The normal distribution is the usual bell shaped curve, but the uniform distribution is the rectangular or box shaped graph. The two distributions are very different.

The uniform distribution has the property that all subintervals of the same length inside the interval 0 to 9 have the same probability of occurrence no matter where they are located. This means that the uniform distribution on the interval from 0 to 9 could be represented on the computer by selecting random numbers from 0 to 9. Since all numbers from 0 to 9 would be equally likely to be chosen, we say we are dealing with a uniform (equally likely) probability distribution. Note that when we say we are selecting random numbers from 0 to 9, we do not just mean whole numbers or integers; we mean real numbers in decimal form such as 2.413912, and so forth.

Because the interval from 0 to 9 is 9 units long and because the total area under the probability graph must be 1 (why?) then the height of the uniform probability graph must be 1/9. The mean of the uniform distribution on the interval from 0 to 9 is the balance point. Looking at the Figure, it is fairly clear that the mean is 4.5. Using advanced methods of statistics, it can be shown that for the uniform probability distribution x between 0 and 9

$$\mu = 4.5 \text{ and } \sigma = 3\sqrt{3}/2 \approx 2.598$$

The figure shows us that the uniform x distribution and the normal distribution are quite different. However, using the computer we will construct one hundred sample means $\bar{x}$ from the x distribution using a sample size of n = 40. We can vary the sample size n according to how many columns we use in RANDOM command.

RANDOM 100 C1-C40;
UNIFORM a = 0 to b = 9.

We will see that even though the uniform distribution is very different from the normal distribution the histogram of the sample means is somewhat bell shaped. Looking at the DESCRIBE command, we will also see that the mean of the $\bar{x}$ distribution is close to the predicted mean of 4.5 and that the standard deviation is close to $\sigma/\sqrt{n}$ or $2.598/\sqrt{40}$ or 0.411.

Example

The following MINITAB program will draw 100 random samples of size 40 from the uniform distribution on the interval from 0 to 9. We put the data into 40 columns. Then we take the mean of each of the 100 rows (40 columns across) and store the result in C50. To do this, we use the RMEAN C1-C40 put into C50 command. Next we DESCRIBE C50 to look at the mean and standard deviation of the distribution of sample means. Finally we look at a histogram of the sample means in C50.

```
MTB > RANDOM 100 C1-C40;
SUBC> UNIFORM a = 0 to b = 9.
MTB > # Take the mean of the 40 data
  > in each row
MTB > # Put the means in C50
MTB > RMEAN C1-C40 put into C50
MTB > DESCRIBE C50
```

```
         N    MEAN  MEDIAN  TRMEAN
C50    100  4.5559  4.5886  4.5531

     STDEV  SEMEAN     MIN     MAX
C50 0.4044  0.0404  3.7003  5.4569

        Q1      Q3
C50 4.1897  4.8487

MTB > GSTD
MTB > HISTOGRAM C50

Histogram of C50   N = 100

Midpoint  Count
   3.8       6  ******
   4.0      10  **********
   4.2      12  ************
   4.4       8  ********
   4.6      25  *************************
   4.8      21  *********************
   5.0      11  ***********
   5.2       2  **
   5.4       5  *****
```

Note the MEAN and STDEV are very close to the values predicted by the Central Limit Theorem.

The histogram for this sample does not appear very similar to a normal distribution. Let's try another sample.

```
MTB > RANDOM 100 C1-C40;
SUBC> UNIFORM a = 0 to b = 9.
MTB > RMEAN C1-C40 put into C50
MTB > DESCRIBE C50

         N    MEAN  MEDIAN  TRMEAN
C50    100  4.5265  4.5274  4.5284

     STDEV  SEMEAN     MIN     MAX
C50 0.4021  0.0402  3.4106  5.4145

        Q1      Q3
C50 4.2179  4.7927

MTB > GSTD
MTB > HISTOGRAM C50

Histogram of C50   N = 100

Midpoint  Count
```

This histogram looks more like a normal distribution. You will get slightly different results each time you draw 100 samples.

```
3.4     1 *
3.6     0
3.8     5 *****
4.0     9 *********
4.2    14 **************
4.4    18 ******************
4.6    19 *******************
4.8    18 ******************
5.0     7 *******
5.2     6 ******
5.4     3 ***
```

The number of samples used is determined by K, and the size of the samples is determined by the number of columns in the command

RANDOM K C1-CN;
UNIFORM a = 0 to b = 9.

Be sure that when you take the RMEAN of the rows, you use the same number of columns as you used in the random command

RMEAN C1-CN put into C

Then use the DESCRIBE and HISTOGRAM commands on the column C where you put the means.

You can sample from a variety of distributions, some of which were listed under the RANDOM command in the Chapter 2 Command Summary.

Lab Activities for Central Limit Theorem

1. Repeat the experiment of Example 1. That is, draw 100 random samples of size 40 each from the uniform probability distribution between 0 and 9. Then take the means of each of these samples and put the results in C50. Use the commands

```
RANDOM 100 C1-C40
  UNIFORM a = 0 to b = 9
RMEAN C1-C40 put into C50
```

Next use DESCRIBE on C50. How does the mean and standard deviation of the distribution of sample means compare to those predicted by the Central Limit Theorem? Use HISTOGRAM C50 to draw a histogram of the distribution of sample means. How does it compare to a normal curve? (Note: in the student edition of MINITAB, the worksheet may not be large enough to accommodate this project. Change the sample size to 15 - that is use C1-C15 in the RANDOM and RMEAN commands.)

2. Next take 100 random samples of size 20 from the uniform probability distribution between 0 and 9. To do this, use only 20 columns in the RANDOM and RMEAN commands. Again put the means in C50, use DESCRIBE and HISTOGRAM on C50 and comment on the results. How do these results compare to those in problem 1? How do the standard deviations compare? (Note: in the student edition of MINITAB, the worksheet may not be large enough to accommodate this project. Change the sample size to 10)

CHAPTER 8 ESTIMATION

CONFIDENCE INTERVALS FOR A MEAN OR FOR A PROPORTION (Sections 8.1, 8.2, 8.3)

Student's t Distribution

In Section 8.1 of *Understandable Statistics* confidence intervals for μ using large samples are presented In Section 8.2 the Student's t distribution is introduced and confidence intervals for μ using small samples are discussed. If the sample size n is small (n < 30) then the $\bar{x}$ distribution follows the Student's t distribution with degrees of freedom (n - 1).

$$t = \frac{\bar{x} - \mu}{\frac{s}{\sqrt{n}}}$$

There is a different Student's t distribution for every degree of freedom. MINITAB includes the Student's t distribution in its library of probability distributions. You may use the RANDOM, PDF, CDF, INVCDF commands with the Student's t distribution as the specified distribution.

Menu selection: **Calc ➤ Probability Distribution ➤ t**

Dialogue Box Responses

Select from Prob Density (PDF), Cumulative Probability (CDF), Inverse Cumulative Probability (INVCDF)

Degrees of Freedom: enter value

Input Column: Column containing values for which you wish to compute the probability and optional storage column

Input Constant: If you want the probability of just one value, use a constant rather than and entire column. Designate optional storage constant or column.

You can graph different t-distributions by using **➤Graph ➤ Plot**. Follow steps similar to those given in Chapter 6 for graphing a normal distribution. The Student's t distribution is symmetric and centered at 0. Select X values from about -4 to 4 in increments of 0.10 and place the values in a column, say C1. Then use **Calc ➤ Probability Distribution ➤ t** with Probability Density to generate a column, say C2 of Y values.

Student's t Distribution with 10 Degrees of Freedom

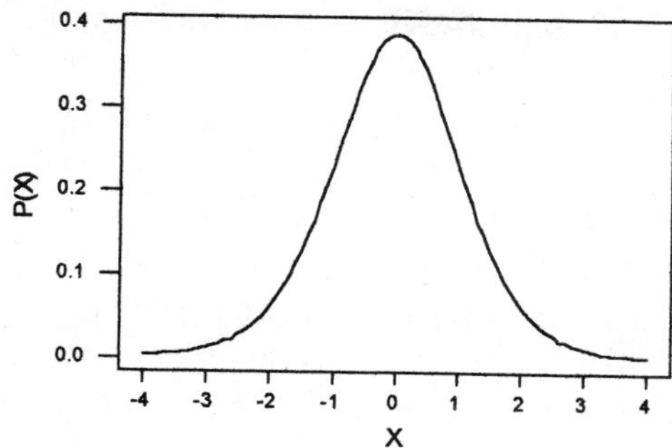

Confidence Intervals for Means

Confidence intervals for μ depend on the sample size n, and knowledge about the standard deviation σ. For small samples we assume the x distribution is approximately normal (mound shaped and symmetric). The relationship between confidence intervals for μ, sample size, and knowledge about σ are shown in the table below.

	Confidence Interval
Large samples σ known σ estimated by s Small samples σ known	$\overline{x} - z(\sigma/\sqrt{n})$ to $\overline{x} + z(\sigma/\sqrt{n})$
Small Samples σ unknown	$\overline{x} - t(s/\sqrt{n})$ to $\overline{x} + t(s/\sqrt{n})$

In MINITAB we can generate confidence intervals for μ by using the menu selections

➤ Stat ➤Basic Statistics ➤1 sample z

Dialogue Box Responses

Variables: Designate the column number C# containing the data

Confidence Interval: Give the level, such as 90%

Test Mean: Leave black at this time. We will use this option in Chapter 9

Sigma: Enter the value of sigma, or of the sample standard deviation for large samples. Note that MINITAB requires knowledge of σ before you can use the normal distribution for confidence intervals

Graphs: You can select from histogram, dot plot, or box plot of sample data

➤ Stat ➤Basic Statistics ➤1 sample t

Dialogue Box Responses

Variables: Designate the column number C# containing the data

Confidence Interval: Give the level, such as 90%

Test Mean: Leave black at this time. We will use this option in Chapter 9

Graphs: You can select from histogram, dot plot, or box plot of sample data

Example

The manager of First National Bank wishes to know the average waiting times for student loan application action. A random sample of 20 applications showed the waiting times from application submission (in days) to be

3	7	8	24	6	9	12	25	18	17
4	32	15	16	21	14	12	5	18	16

Find a 90% confidence interval for the population mean of waiting times.

In this example we have a small sample and σ is not known. We need to use the Student's t distribution. Enter the data into column C1 and name the column Days. Use the menu selection ➤ **Stat** ➤**Basic Statistics** ➤**1 sample t**.

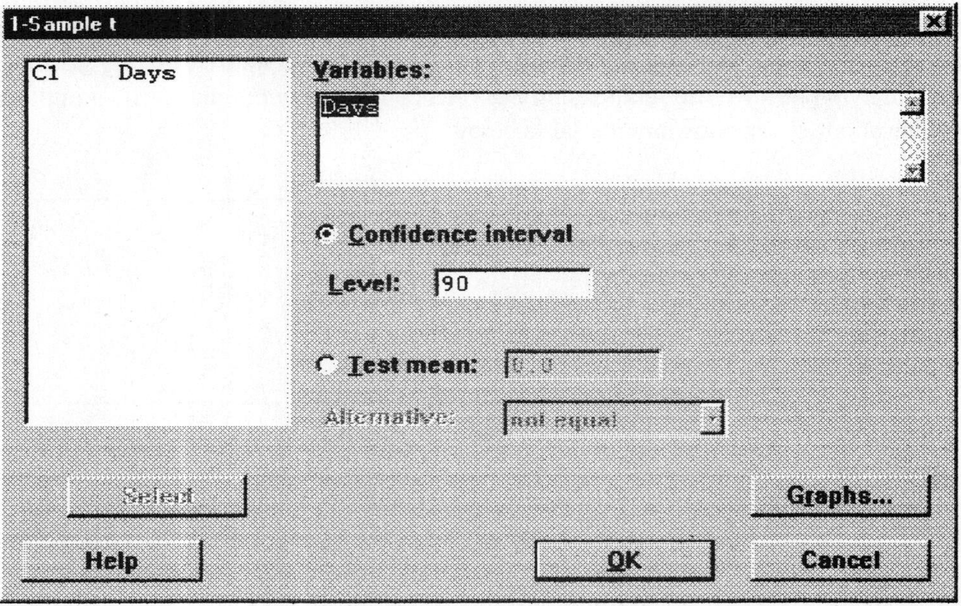

The results are

T Confidence Intervals

```
Variable      N       Mean     StDev  SE Mean        90.0 % CI
Days         20      14.10      7.70     1.72  (   11.12,    17.08)
```

Confidence Intervals for Proportions

This option is in more recent versions of MINITAB, and is in Version 12.

Menu selection

➤ **Stat** ➤ **Basic Statistics** ➤ **1-proportion**

Dialogue Box Responses

Select the option of Summarized Data.

Number of Trials: Enter value (n in *Understandable Statistics*)

Number of Successes: Enter value (r in *Understandable Statistics*)

Click on [Options]; Enter confidence level and click on use normal distribution.

In Chapter 9 we will see how to interpret the results from test.

Example

The public television station BPBS wants to find the percent of its viewing population who give donations to the station. A random sample of 300 viewers were surveyed and it was found that 123 made contributions to the station. Find a 95% confidence interval for the probability that a viewer of BPBS selected at random contributes to the station.

Use the menu selection ➤ **Stat** ➤ **Basic Statistics** ➤ **1-proportion**. Use 300 for number of trials and 123 for number of successes. Click on [Options]. Enter 95 for the confidence level.

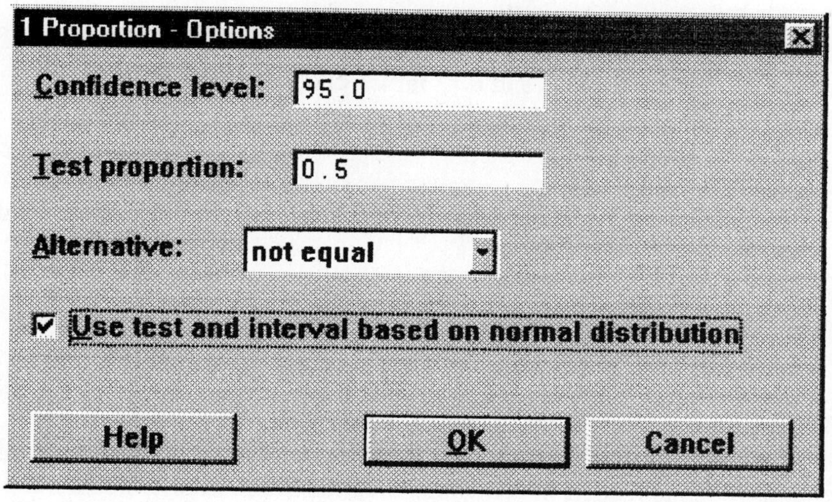

1 Proportion - Options

Confidence level: 95.0

Test proportion: 0.5

Alternative: not equal

☑ Use test and interval based on normal distribution

Help OK Cancel

The results follow

Test and Confidence Interval for One Proportion

```
Test of p = 0.5 vs p not = 0.5

Sample      X     N  Sample p         95.0 % CI          Z-Value  P-Value
1         123   300  0.410000  (0.354345, 0.465655)       -3.12    0.002
```

The output regarding test of p, z-value, and p-value will be discussed in Chapter 9.

CONFIDENCE INTERVALS FOR DIFFERENCE OF MEANS OR DIFFERENCE OF PROPORTION

In MINITAB, confidence intervals for difference of means and difference of proportions are included in the menu selections for tests of hypothesis for difference of means and tests of hypothesis for difference of proportions respectively. These menu selections with their dialogue boxes will be discussed in Chapter 9.

Lab Activities for Confidence Intervals for the Mean or a Proportion

1. Snow King Ski resorts is considering opening a downhill ski slope in Montana. To determine if there would be an adequate snow base in November in the particular region under consideration, they studied snow fall records for the area over the last 100 years. They took a random sample of 15 years. The snow fall during November in inches for the sample years was (in inches)

   26 35 42 18 29 42 28 35
   47 29 38 27 21 35 30

   a) To find a confidence interval for μ do we use a normal distribution or a student's t distribution?
   b) Find a 90% confidence interval for the mean snow fall.
   c) Find a 95% confidence interval for the mean snow fall.
   d) Compare the intervals of parts (b) and (c). Which one is narrower? Why would you expect this?

2. Consider the snow fall data of problem 1. Suppose you knew that the snow fall in the region under consideration for the ski area in Montana (see problem 1) had a population standard deviation of 8 inches.

   a) Since you know σ, (and the distribution of snow fall is assumed to be approximately normal) do you use the normal distribution or student's t for confidence intervals?
   b) Find a 90% confidence interval for the mean snow fall.
   c) Find a 95% confidence interval for the mean snow fall.
   d Compare the respective confidence intervals created in problem 1 and in this problem. Of the 95% intervals, which is longer, the one using the t distribution or the one using the normal distribution? Why would you expect this result?

3. Retrieve the worksheet DISN.mtp from the DATA DISK. This worksheet contains the number of shares of Disney Stock (in hundreds of shares) sold for a random sample of 60 trading days in 1993 and 1994. The data is in column C1.

   Use the sample standard deviation computed with menu options ➤ Stat ➤Basic Statistics ➤Display Descriptive Statistics as the value of σ. You will need to compute this value first, and then enter it in as a number in the dialog box for 1-sample z.

   a) Find a 99% confidence interval for the population mean volume.
   b) Find a 95% confidence interval for the population mean volume
   c) Find a 90% confidence interval for the population mean volume

d) Find an 85% confidence interval for the population mean volume.
e) What do you notice about the lengths of the intervals as the confidence level decreases?

4. There are many types of errors that will cause a computer program to terminate or give incorrect results. One type of error is punctuation. For instance, if a comma is inserted in the wrong place, the program might not run. A study of programs written by students in a beginning programming course showed that 75 out of 300 errors selected at random were punctuation errors. Find a 99% confidence interval for the proportion of errors made by beginning programming students that are punctuation errors. Next find a 90% confidence interval. Is this interval longer or shorter?

5. Sam decided to do a statistics project to determine a 90% confidence interval for the probability that a student at West Plains College eats lunch in the school cafeteria. He surveyed a random sample of 12 students and found that 9 ate lunch in the cafeteria. Can Sam use the program to find a confidence interval for the population proportion of students eating in the cafeteria? Why or why not? Try the program with N = 12 and R = 9. What happens? What should Sam do to complete his project?

COMMAND SUMMARY

Probability distribution subcommand

T with df = K is the subcommand that calls up the Student's t distribution with specified degrees of freedom K. This subcommand may be used with RANDOM, PDF, CDF, INVCDF

WINDOWS menu selection: **Calc ➤ Probability Distribution ➤ t**

In the dialog box select PDF, CDF, or Inverse; Enter the degrees of freedom.

To generate confidence intervals

ZINTERVAL [K% confidence] σ = K on C...C generates a confidence interval for μ using the normal distribution. You must enter a value for σ, either actual or estimated. A separate interval is given for data in each column. If K is not specified, a 95% confidence interval will be given.

WINDOWS menu selection: **Stat➤Basic Statistics➤1 sample z**

In the dialog box select confidence interval and enter the confidence level

TINTERVAL [K% confidence] for C...C generates a confidence interval for μ using the Student's t distribution. It automatically computes stdev s from the data as well as the number of degrees of freedom. If K is not specified, a 95% confidence interval is given.

WINDOWS menu selection: **Stat➤Basic Statistics➤1 sample t**

In the dialog box select confidence interval and enter the confidence level

PONE [N R] with **subcommand Confidence C** generates a confidence interval for one proporiton.

WINDOWS menu selection **Stat➤Basic Statistics➤1 proportion**

CHAPTER 9 HYPOTHESIS TESTING

TESTING A SINGLE POPULATION MEAN OR PROPOTION

Chapter 9 of *Understandable Statistics* introduces tests of hypotheses. Tests involving a single mean are found in Sections 9.2 (large samples) and 9.4 (small samples). In MINITAB, the user concludes the test by comparing the P-value of the test statistic to the level of significance α. The method of using P-values to conclude tests of hypotheses is explained in Section 9.3 of *Understandable Statistics*. Section 9.5 of *Understandable Statistics* discusses tests of a single proportion.

For tests of the mean with large samples use
➤ **Stat** ➤**Basic Statistics** ➤ **1 sample z**
Dialogue Box Responses
Variable: Enter column number where data is located
Select Test Mean. Enter the value of K for the null hypothesis
H_0: $\mu = K$
Alternative: Scroll to the relation in the alternate hypothesis
H_1: $\mu \neq K$ (not equal)
H_1: $\mu > K$ (greater than)
H_1: $\mu < K$ (less than)
Sigma: Use the value of sigma, or estimate it with the sample standard deviation s. Recall that you can use the ➤**Stat** ➤**Basic Statistics** ➤**Display Descriptive Statistics** menu choices to find the value of s

For tests of the mean with small samples use
➤ **Stat** ➤**Basic Statistics** ➤ **1 sample t**
Dialogue Box Responses
Variable: Enter column number where data is located
Select Test Mean. Enter the value of K for the null hypothesis
H_0: $\mu = K$
Alternative: Scroll to the relation in the alternate hypothesis
H_1: $\mu \neq K$ (not equal)
H_1: $\mu > K$ (greater than)
H_1: $\mu < K$ (less than)

For tests of a single proportion use
➤ **Stat** ➤**Basic Statistics** ➤ **1 proportion**
Dialogue Box Responses
Select on Summarized data; enter the number of trials and the number of successes
Click on [Options].
Confidence Level: Enter a value such as 95
Test proportion: Enter the value of K where
H_0: $p = K$
Alternative: Scroll to the relation in the alternate hypothesis
H_1: $p \neq K$ (not equal)
H_1: $p > K$ (greater than)
H_1: $p < K$ (less than)

Both the Z-sample and T-sample operate on data in a column. They each compute the sample mean $\bar{x}$. The Z-sample converts the sample mean $\bar{x}$ to a z value while the T-sample converts $\bar{x}$ to t using the respective formulas

$$z = \frac{\bar{x} - \mu}{\sigma/\sqrt{n}} \qquad t = \frac{\bar{x} - \mu}{s/\sqrt{n}}$$

The test of 1 proportion converts the sample proportion $\hat{p} = r/n$ to a z value using the formula

$$z = \frac{(r/n) - p}{\sqrt{(P(1-p)/n)}}$$

The tests also give the P-value of the sample statistic $\bar{x}$. The user can then compare the P-value to α, the level of significance of the test. If

P-value $\leq \alpha$ we reject the null hypothesis
P-value $> \alpha$ we do not reject the null hypothesis

Example

Many times patients visit a health clinic because they are ill. A random sample of 12 patients visiting a health clinic had temperatures (in °F)

97.4	99.3	99.0	100.0	98.6
97.1	100.2	98.9	100.2	98.5
98.8	97.3			

Dr. Tafoya believes that patients visiting a health clinic have a higher temperature than normal. The normal temperature is 98.6 degrees. Test the claim at the $\alpha = 0.01$ level of significance.

In this case, we have a small sample and do not know σ. We need a t-test. Enter the data in C1 and name the column Temp. Then select ➤ **Stat** ➤**Basic Statistics** ➤ **1 sample t**

Use 98.6 as the value for Test mean. Scroll to greater than for Alternative.

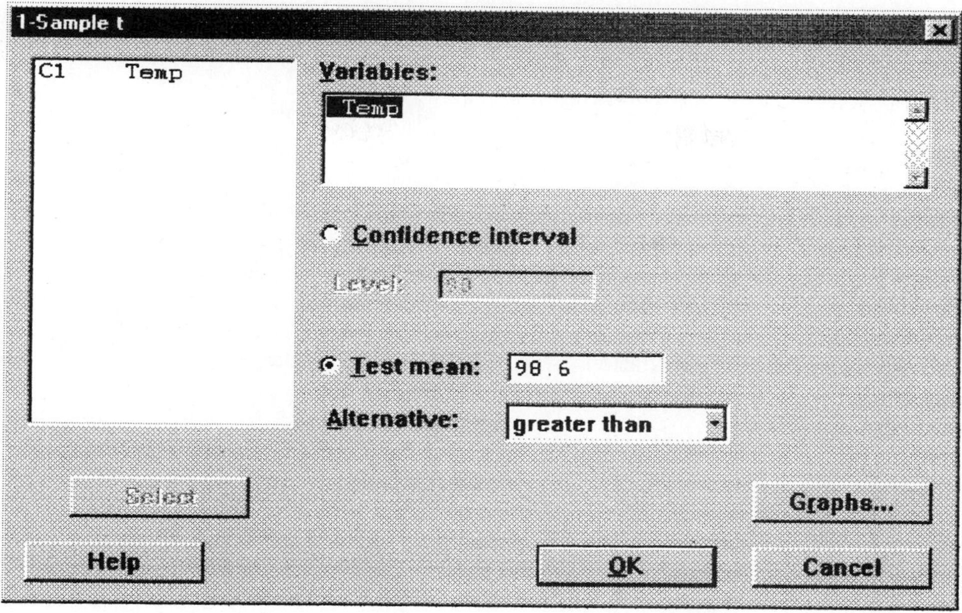

The results follow

T-Test of the Mean

```
Test of mu = 98.600 vs mu > 98.600

Variable      N      Mean     StDev   SE Mean       T        P
Temp         12    98.775     1.082     0.312    0.56     0.29
```

Recall that SE Mean is the value of $\frac{s}{\sqrt{n}}$.

Lab Activities for Testing a Single Population Mean or Single Proportion

1. A new catch and release policy was established for a river in Pennsylvania. Prior to the new policy, the average number of fish caught per fisherman hour was 2.8. Two years after the policy went into effect a random sample of 12 fisherman hours showed the following catches per hour.

   3.2 1.1 4.6 3.2 2.3 2.5
   1.6 2.2 3.7 2.6 3.1 3.4

   Test the claim that the per hour catch has increased at the 0.05 level of significance.
   a) Decide whether to use the Z-sample or T-sample menu choices. What is the value of μ in the null hypothesis?
   b) What is the choice ALTERNATIVE?
   c) Compare the P-value of the test statistic to the level of significance α. Do we reject the null hypothesis or not?

2. Open or retrieve the worksheet **MPGAL.mpt** from the Data Disk. The data in column C1 of this worksheet represent the miles per gallon gasoline consumption (highway) for a random sample of

55 makes and models of passenger cars (source: Environmental Protection Agency).

30	27	22	25	24	25	24	15
35	35	33	52	49	10	27	18
20	23	24	25	30	24	24	24
18	20	25	27	24	32	29	27
24	27	26	25	24	28	33	30
13	13	21	28	37	35	32	33
29	31	28	28	25	29	31	

Test the hypothesis that the population mean miles per gallon gasoline consumption for such cars is greater than 25 mpg.

a) Do we know σ for the mpg consumption? Can we estimate σ by s, the sample standard deviation? Should we use the Z-sample or T-sample menu choice? What is the value of μ in the null hypothesis?
b) If we estimate σ by s, we need to instruct MINITAB to find the stdev of the data before we use Z-sample. Use **➤Stat ➤Basic Statistics ➤Display Descriptive Statistics** to find s.
c) What is the alternate hypothesis?
d) Look at the P-value in the output. Compare it to α. Do we reject the null hypothesis or not?
e) Using the same data, test the claim that the average mpg for these cars is not equal to 25. How has the P-value changed? Compare the new P-value to α. Do we reject the null hypothesis or not?

3. Open or retrieve the worksheet **WOLF.mpt** from the Data Disk. The data in column C1 of this worksheet represent the number of wolf pups per den from a sample of 16 wolf dens (source: *The Wolf in the Southwest: The making of an Endangered Species* by D.E. Brown, University of Arizona Press).

5.00	8.00	7.00	5.00	3.00	4.00	3.00	9.00
5.00	8.00	5.00	6.00	5.00	6.00	4.00	7.00

Test the claim that the population mean number of wolf pups in a den is greater than 5.4.

4. Jones Computer Security is testing a new security device which is believed to decrease the incidence of computer "break ins." Without this device, the computer security test team can break security 47% of the time. With the device in place, the test team made 400 attempts and were successful 82 times. Select an appropriate program from the HYPOTHESIS TESTING menu and test the claim that the device reduces the proportion of successful break ins. Use alpha = 0.05 and note the P-value. Does the test conclusion change for alpha = 0.01?

TESTS INVOLVING PAIRED DIFFERENCES (DEPENDENT SAMPLES)

The test for difference of means, dependent samples is presented in Section 9.6 of *Understandable Statistics*. Dependent samples arise from before and after studies, some studies of data taken from the same subjects, and some studies on identical twins.

To perform a paired difference test, we put our paired data into two columns. New in later versions of MINITAB including the Student Edition 12 for Windows is a menu item for testing data pairs. Select

➤Stat ➤Basic Statistics ➤Paired-t

Dialogue Box Responses

First Sample: column number where before data is located

Second Sample: column number where after data is located

Click [Options]

Confidence Level: Enter a value such as 95

Test mean: Leave as default 0.0

Alternative: Scroll to not equal, greater than, or less than as appropriate

H_1: $\mu_d \neq 0$ (not equal)

H_1: $\mu_d > 0$ (greater than)

H_1: $\mu_d < 0$ (less than)

Example

Promoters of a state lottery decided to advertise the lottery heavily on television for one week during the middle of one of the lottery games. To see if the advertising improved ticket sales, they surveyed a random sample of 8 ticket outlets and recorded weekly sales for one week before the television campaign and for one week after the campaign. The results follow (in ticket sales) where B standard for before and A for after the advertising campaign.

B:	3201	4529	1425	1272	1784	1733	2563	3129
A:	3762	4851	1202	1131	2172	1802	2492	3151

Test the claim that the television campaign increased lottery ticket sales at the 0.05 level of significance.

We want to test to see if D = B - A is less than zero since we are testing the claim that the lottery ticket sales are greater after the television campaign. We will put the before data in C1, the after data in C2. Select **➤Stat ➤Basic Statistics ➤Paired-t**. Use less than for Alternative.95.0

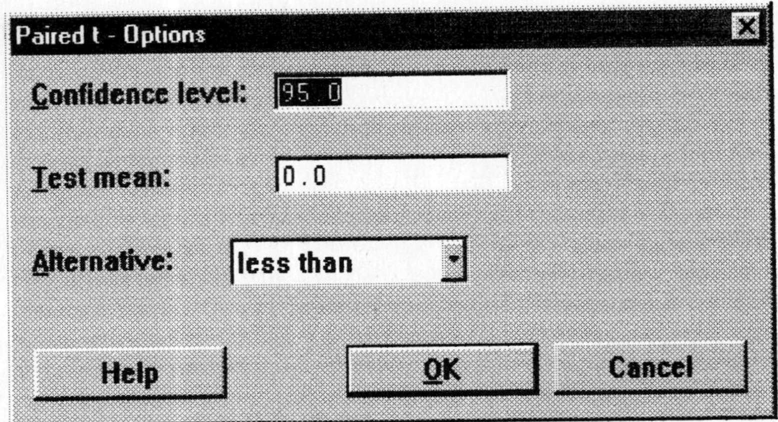

The results are

Paired T-Test and Confidence Interval

```
Paired T for Before - After

                N        Mean      StDev    SE Mean
Before          8        2455       1118        395
After           8        2570       1291        456
Difference      8      -115.9      278.1       98.3

95% CI for mean difference: (-348.5, 116.8)
T-Test of mean difference = 0 (vs < 0): T-Value = -1.18
  P-Value = 0.139
```

Since the P value 0.139 is larger than the level of significance of 0.05, we do not reject the null hypothesis.

Lab Activities using TESTS INVOLVING PAIRED DIFFERENCES (DEPENDENT SAMPLES)

1. Open or retrieve the worksheet **FSALARY.mtp** from the DATA DISK. The data are pairs of values where the entry in C1 represents average salary ($1000/yr) for male faculty members at an institution and C2 represents the average salary for female faculty members ($1000/yr) at the same institution. A random sample of 22 U.S. colleges and universities was used (source: Academe, Bulletin of the American Association of University Professors).

   (34.5, 33.9) (30.5, 31.2) (35.1, 35.0) (35.7, 34.2) (31.5, 32.4)
   (34.4, 34.1) (32.1, 32.7) (30.7, 29.9) (33.7, 31.2) (35.3, 35.5)
   (30.7, 30.2) (34.2, 34.8) (39.6, 38.7) (30.5, 30.0) (33.8, 33.8)
   (31.7, 32.4) (32.8, 31.7) (38.5, 38.9) (40.5, 41.2) (25.3, 25.5)
   (28.6, 28.0) (35.8, 35.1)

   a) The data is in C1 and C2
   b) Use the ➤**Stat** ➤**Basic Statistics** ➤**Paired-t** menu to test the hypothesis that there is a difference in salary. What is the P value of the sample test statistic? Do we reject or fail to

reject the null hypothesis at the 5% level of significance? What about at the 1% level of significance?

c) Use the ➤**Stat** ➤**Basic Statistics** ➤**Paired-t** menu to test the hypothesis that female faculty members have a lower average salary than male faculty members. What is the test conclusion at the 5% level of significance? At the 1% level of significance?

2. An audiologist is conducting a study on noise and stress. Twelve subjects selected at random were given a stress test in a room that was quiet. Then the same subjects were given another stress test, this time in a room with high pitch background noise. The results of the stress tests were scores 1 through 20 with 20 indicating the greatest stress. The results follow where **B** represents the score of the test administered in the quiet room and A represents the scores of the test administered in the room with the high pitch background noise.

Subject		1	2	4	5	6	7	8	9	10	11	12
	B	13	12	16	19	7	13	9	15	17	6	14
	A	18	15	14	18	10	12	11	14	17	8	16

Test the hypothesis that the stress level was greater during exposure to high pitch background noise. Look at the P-value. Should you reject the null hypothesis at the 1% level of significance? At the 5% level?

TESTS OF DIFFERENCE OF MEANS (INDEPENDENT SAMPLES)

Tests of difference of means, independent samples are presented in Sections 9.7 *Understandable Statistics*. We consider the $\bar{x}_1 - \bar{x}_2$ distribution. The null hypothesis is that there is no difference between means so $H_0{:}\mu_1 = \mu_2$.

Large Samples

MINITAB has a slightly different approach to testing difference of means with large samples (each sample size 30 or more) than that shown in *Understandable Statistics*. In MINITAB the Student's t distribution is used instead of the normal distribution. The degrees of freedom used by MINITAB for this application of the t distribution is at least as large as the smaller sample. Therefore we have degrees of freedom at 30 or more. In such cases the normal and student's t distribution give reasonably similar results. However, the results will not be exactly the same.

The menu choice MINITAB uses to test the difference of means is **STAT➤ Basic Statistics➤ 2 sample t**. The null hypothesis is always $H_0{:}\mu_1 = \mu_2$. The alternate hypothesis $H_1{:}\mu_1 \neq \mu_2$. corresponds to the choice not equal. To do a left tail or right tail test, you need to use the choice of less than for ALTERNATIVE on a left tail test and greater than for ALTERNATIVE on a right tail test.

WINDOWS menu selection: **STAT➤ Basic Statistics➤ 2 sample t**

Dialogue Box Responses

Select Samples in Different Columns and enter the C# for the columns containing the data
Alternative: Scroll to the appropriate choice
Confidence Level: Enter a value such as 95
Assume equal variances: Do **not** select for large samples

Small Samples

To do a test of difference of sample means with small samples with the assumption that the samples come from populations with the same standard deviation, we use the **STAT➤ Basic Statistics➤ 2 sample t** menu selection with **Assume equal variances** checked. When we check that equal variances are assumed, MINITAB automataically pools the standard deviations.

➤ STAT➤ Basic Statistic s➤ 2 sample t

Dialogue Box Responses
Select Samples in Different Columns and enter the C# for the columns containing the data
Alternative: Scroll to the appropriate choice
Confidence Level: Enter a value such as 95
Assume equal variances: Check this item so that the pooled standard deviation is used

Example

Sellers of microwave French fry cookers claim that their process saves cooking time. McDougle Fast Food Chain is considering the purchase of these new cookers, but wants to test the claim. Six batches of French fries were cooked in the traditional way.
Cooking times (in minutes) are

15 17 14 15 16 13

Six batches of French fries of the same weight were cooked using the new microwave cooker. These cooking times (in minutes) are

11 14 12 10 11 15

Test the claim that the microwave process takes less time. Use $\alpha = 0.05$.

Under the assumption that the distribution of cooking times for both methods are approximately normal, and that $\sigma_1 = \sigma_2$ we use the ➤ **STAT**➤ **Basic Statistic s**➤ **2 sample t** menu choices with the assumption of equal variances checked. We are testing the claim that the mean cooking time of the second sample is less than that of the first sample, so our alternate hypothesis will be H_1: $\mu_1 > \mu_2$. We will use a right tail test and scroll to greater than for ALTERNATIVE

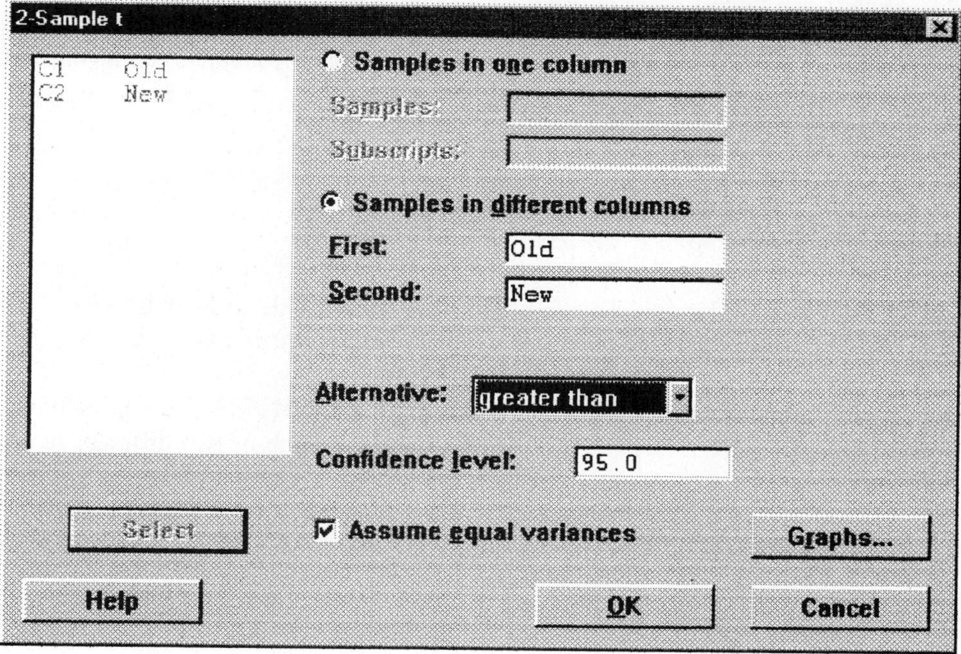

The results are

Two Sample T-Test and Confidence Interval

```
Two sample T for Old vs New

      N      Mean      StDev    SE Mean
Old   6      15.00      1.41       0.58
New   6      12.17      1.94       0.79

95% CI for mu Old - mu New: ( 0.65,  5.02)
T-Test mu Old = mu New (vs >): T = 2.89   P = 0.0081   DF = 10
   Both use Pooled StDev = 1.70
```

We see that the P-value of the test is 0.0081. Since the P-value is less than $\alpha = 0.05$, we reject the null hypothesis and conclude that the microwave method takes less time to cook French fries.

Lab Activities Using Difference of Means (Independent Samples)

1. Calm Cough Medicine is testing a new ingredient to see if its addition will lengthen the effective cough relief time of a single does. A random sample of 15 doses of the standard medicine were tested and the effective relief times were (in minutes):

   42 35 40 32 30 26 51 39 33 28
   37 22 36 33 41

   A random sample of 20 doses were tested when the new ingredient was added. The effective relief times were (in minutes):

   43 51 35 49 32 29 42 38 45 74
   31 31 46 36 33 45 30 32 41 25

   Assume that the standard deviations of the relief times are equal for the two populations. Test the claim that the effective relief time is longer when the new ingredient is added. Use $\alpha = 0.01$.

2. Open or retrieve the worksheet **RABIES.mtp** from the DATA DISK. The data represent number of cases of red fox rabies for a random sample of 16 areas in each of two different regions of southern German.

   NUMBER CASES IN REGION 1

   10 2 2 5 3 4 3 3 4 0 2 6 4 8 7 4

   NUMBER CASES IN REGION 2

   1 1 2 1 3 9 2 2 4 5 4 2 2 0 0 2

   Test the hypothesis that the average number of cases in Region 1 is greater than the average number of cases in Region 2. Use a 1% level of significance.

3. Open or retrieve the MINITAB worksheet **PETAL.mtp** from the DATA DISK. The data represent the petal length (cm) for a random sample of 35 Iris Virginica and for a random sample of 38 Iris Setosa (source: Anderson, E., Bull. Amer. Iris Soc).

PETAL LENGTH (C.M.) IRIS VIRGINICA

5.1 5.8 6.3 6.1 5.1 5.5 5.3 5.5 6.9 5.0 4.9 6.0 4.8 6.1 5.6 5.1
5.6 4.8 5.4 5.1 5.1 5.9 5.2 5.7 5.4 4.5 6.1 5.3 5.5 6.7 5.7 4.9
4.8 5.8 5.1

PETAL LENGTH (C.M.) IRIS SETOSA

1.5 1.7 1.4 1.5 1.5 1.6 1.4 1.1 1.2 1.4 1.7 1.0 1.7 1.9 1.6 1.4
1.5 1.4 1.2 1.3 1.5 1.3 1.6 1.9 1.4 1.6 1.5 1.4 1.6 1.2 1.9 1.5
1.6 1.4 1.3 1.7 1.5 1.7

Test the hypothesis that the average petal length for the Iris Setosa is shorter than the average petal length for the Iris Virginica.

COMMAND SUMMARY

To test a single mean

ZTEST [μ = K] σ = K, for C...C performs a z-test on the data in each column. If you do not specify μ, it is assumed to be 0. You need to supply a value for σ (either actual, or estimated by the sample standard deviation s of a column in the case of large samples). If the ALTERNATIVE subcommand is not used, a two tail test is conducted.

WINDOWS menu selection: **Stat ➤ Basic Statistics ➤ 1 sample z**
In dialog box select alternate hypothesis, specify the mean for H_0, specify the standard deviation

TTEST [μ = K] on C...C performs a separate t-test on the data of each column. If you do not specify μ, it is assumed to be 0. The computer evaluates s, the sample standard deviation for each column, and uses the computed s value to conduct the test. If the ALTERNATIVE subcommand is not used, a two tail test is conducted.

WINDOWS menu selection: **Stat ➤ Basic Statistics ➤ 1 sample t**
In dialog box select alternate hypothesis, specify the mean for H_0

ALTERNATIVE = K is the subcommand required to conduct a one tail test. If K = -1, then a left tail test is done. If K = 1, then a right tail test is done.

To test a difference of means (independent samples)

TWOSAMPLE [K% confidence] for C C does a two (independent) sample t test and (optionally confidence interval) for data in the two columns listed. The first data set is put into the first column, and the second data set into the second column. Unless the ALTERNATIVE subcommand is used, the alternate hypothesis is assumed to be $H_1: \mu_1 \neq \mu_2$. Samples are assumed to be independent.

ALTERNATIVE = K is the subcommand to change the alternate hypothesis to a left tail test with K = -1 or right tail test with K = 1.

POOLED is the subcommand to be used only when the two samples come from populations with equal standard deviations.

WINDOWS menu selection: **Stat ➤ Basic Statistics ➤ 2 sample t**
In dialog box select alternate hypothesis, specify the mean for H_0. For large samples do not check assume equal variances. For small samples check assume equal variances

To do a paired difference test (Available on newer versions of MINITAB such as release 12)
PAIRED C C tests for a difference of means in paired (dependent) data and gives a confidence interval if requested.

TEST 0.0 is a subcommand to set the null hypothesis to 0

ALTERNATIVE = K is the subcommand to change the alternate hypothesis to a left tail test with K = -1 or right tail test with K = 1.

WINDOWS menu selection: **Stat ➤ Basic Statistics ➤paired t**
In dialog box select alternate hypothesis, specify the mean for H_0,

CHAPTER 10 REGRESSION AND CORRELATION

SIMPLE LINEAR REGRESSION: TWO VARIABLES

(Sections 10.1, 10.2, 10.3, 10.4 of *Understandable Statistics*)

Chapter 10 of *Understandable Statistics* introduces linear regression. Formulas to find the equation of the least squares line

$$y = a + bx$$

are given in Section 10.2. This section also contains the equation for the standard error of estimate as well as the procedure to find a confidence interval for the predicted value of y. The formula for the correlation coefficient r and coefficient of determination r^2 are given in Section 10.3.

The menu selection **➤Stat ➤Regression ➤ Regression** gives the equation of the least squares line, the value of the standard error of estimate (s = standard error of estimate), the value of the coefficient of determination r^2 (R-sq), as well as several other values such as R-sq adjusted (an unbiased estimate of the population r^2). For simple regression with a response variable and one explanatory variable we can get the value of the Pearson product moment correlation coefficient r by simply taking the square root of R-sq.

The standard deviation, t-ratio and P-values of the coefficients are also given. The P-value is useful for testing the coefficients to see that the population coefficient is not zero (see Section 10.5 *of Understandable Statistics* for a discussion about testing the coefficients). For the time being we will not use these values.

Depending on the amount of output requested (controlled by the options selected under the **[Results]** button) you will also see an analysis of variance chart as well as a table of x and y values with the fitted values y_p and residuals $(y - y_p)$. We will not use the analysis of variance chart in our introduction to regression. However, in more advanced treatments of regression, you will find it useful.

To find the equation of the least squares line and the value of the correlation coefficient, use the menu options

➤Stat ➤Regression ➤ Regression

Dialogue Box Responses

Response: Enter the column number C# of the column containing the responses (that is Y values)

Predictor: Enter the column number C# of the column containing the explanatory variables (that is, X values)

[Graphs]: Do not click on at this time

[Results]: Click on and select the second option, Regression equation, etc

[Options]: We will click on this option when we wish to do predictions for new variables.

[Storage]: Click on and select the fits and residual options if you wish

To graph the scatter plot and show the least squares line on the graph use the menu options

➤Stat ➤Regression ➤Fitted Line Plot

Dialogue Box Responses

Response: List the column number C# of the column containing the Y values

Predictor: List the column number C# of the column containing the X values

Type of Regression model: Select Linear

[Options]: Click on and select Display Prediction Band for a specified confidence level of prediction band. Do not use if you do not want the prediction band.

[Storage]: This button gives you the same storage options as found under regression

To find the value of the correlation coefficient directly and to find its corresponding P-value, use the menu selection

➤Stat ➤Basic Statistics ➤Correlation

Dialogue Box Responses

Variables: List the column number C# of the column containing the X variable and the column number C# of the column containing the Y variable.

Select the P value option.

Example

Merchandise loss due to shoplifting, damage, and other causes is called shrinkage. Shrinkage is a major concern to retailers. The managers of H.R. Merchandise think that there is a relationship between shrinkage and number of clerks on duty. To explore this relationship, a random sample of 7 weeks was selected. During each week the staffing level of sales clerks was kept constant and the dollar value (in hundreds of dollars) of the shrinkage was recorded.

X	10	12	11	15	9	13	8
Y	19	15	20	9	25	12	31 (in hundreds)

Store the value of X in C1 and name C1 as X. Store the values of Y in C2 and name C2 as Y.

Use menu choices to give descriptive statistics regarding the values of X and Y. Use commands to draw an (X,Y) scatter plot and then to find the equation of the regression line. Find the value of the correlation coefficient, and test to see if it is significant.

(a) First we will use **➤Stat ➤Basic Statistics ➤ Display Descriptive Statistics** and each of the columns X, and Y. Note that we select both C1 and C2 in the variables box.

Descriptive Statistics

Variable	N	Mean	Median	TrMean	StDev	SE Mean
X	7	11.143	11.000	11.143	2.410	0.911
Y	7	18.71	19.00	18.71	7.59	2.87

Variable	Minimum	Maximum	Q1	Q3
X	8.000	15.000	9.000	13.000
Y	9.00	31.00	12.00	25.00

(b) Next we will use **➤Stat ➤Regression ➤Fitted Line Plot** to graph the scatter plot and show the least squares line on the graph. We will not use prediction bands.

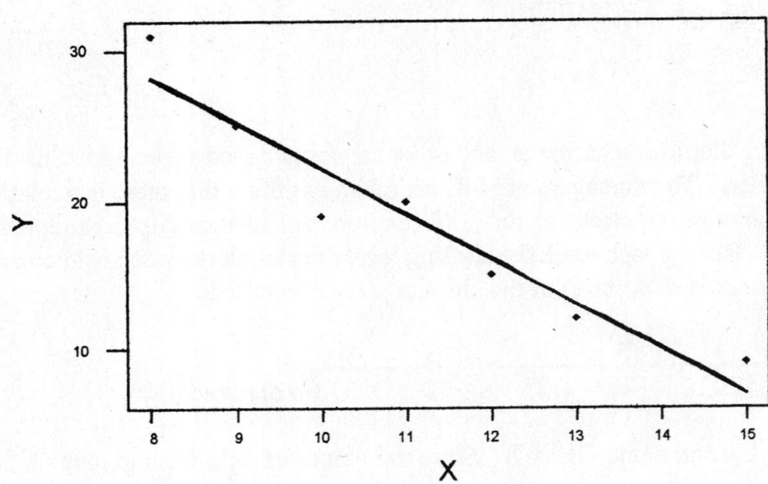

Notice that the equation of the regression line is given on the figure, as well as the value of r^2.

(c) However, to find out more information about the linear regression model, we use the menu selection ➤**Stat** ➤**Regression** ➤ **Regression**. Enter C2 for response and C1 for predictor.

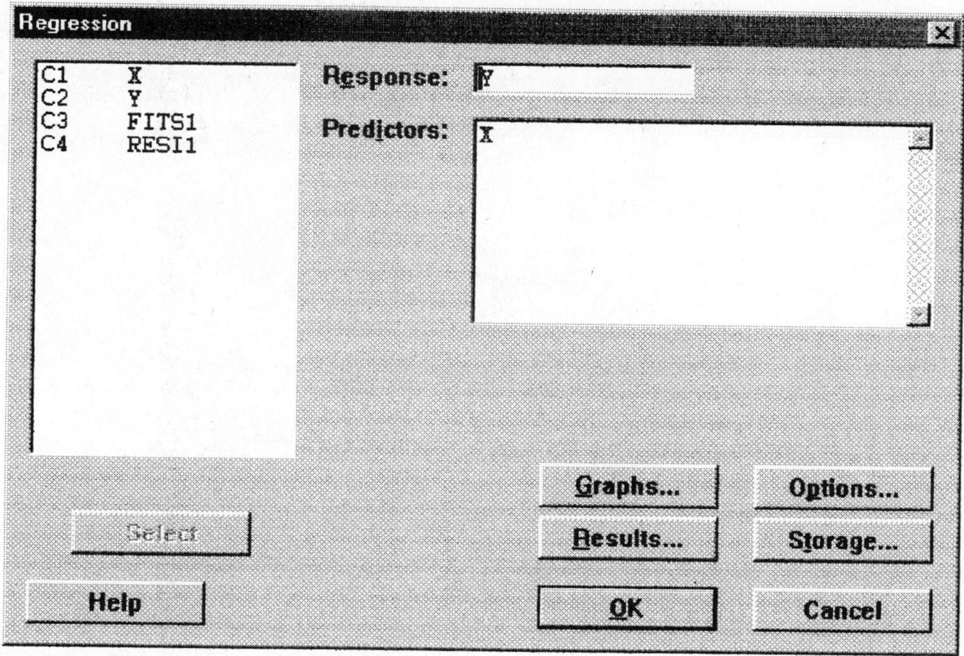

The results follow

Regression

```
The regression equation is
y = 52.5 - 3.03 x

Predictor        Coef        StDev          T        P
Constant       52.508        4.288      12.24    0.000
x             -3.0328       0.3774      -8.04    0.000

S = 2.228        R-Sq = 92.8%      R-Sq(adj) = 91.4%

Analysis of Variance

Source            DF          SS          MS         F        P
Regression         1      320.61      320.61     64.59    0.000
Residual Error     5       24.82        4.96
Total              6      345.43
```

Notice that the regression equation is given as

$$y = 52.5 - 3.03\,x$$

The value of the standard error of estimate S_e is given as S = 2.228. We have the value of r^2, R-square = 92.8%. Find the value of r is found by taking the square root. It is 0.963 or 96.3%.

(d) Next, let's use the prediction option to find the shrinkage when 14 clerks are available.

Use ➤**Stat** ➤**Reression** ➤**Regression**➤**Regression**. Your previous selections should still be listed. Now press [Options]. Enter 14 in the prediction window.

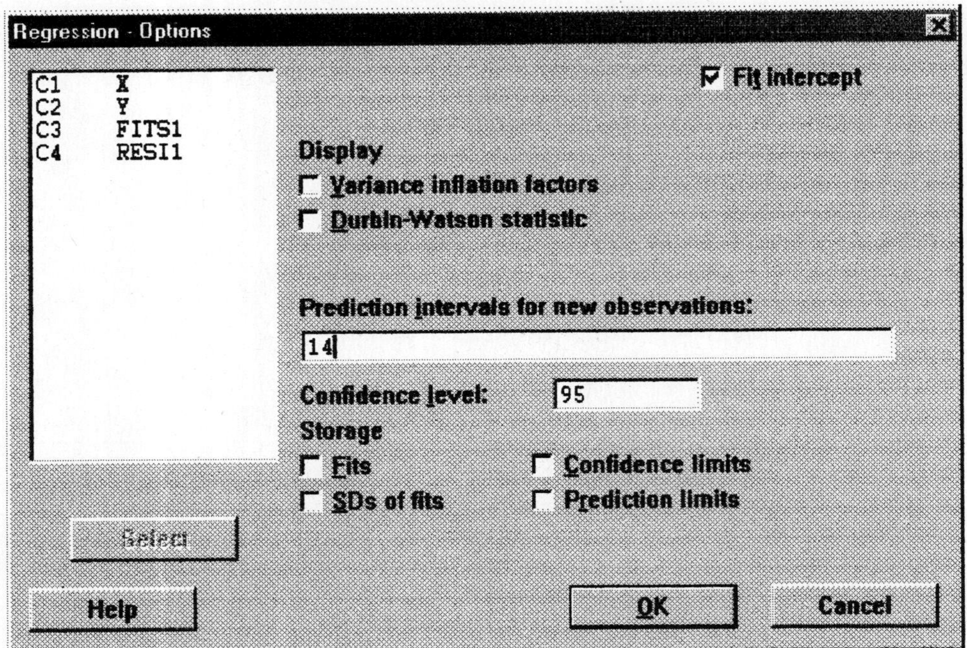

The results follow. The occur after the earlier display.

```
Predicted Values

  Fit     StDev Fit          95.0% CI              95.0% PI
10.049       1.368    (   6.532,   13.566)  (   3.328,   16.771)
```

The predicted value of the shrinkage when 14 clerks are on duty is 10.05 hundred dollars . A 95% prediction interval goes from 3.33 hundred dollars to 16.77 hundred dollars.

(e) Graph a prediction band for predicted values.

Now we use
➤**Stat** ➤**Regression** ➤**Fitted Line Plot** with the [Option] show prediction band selected

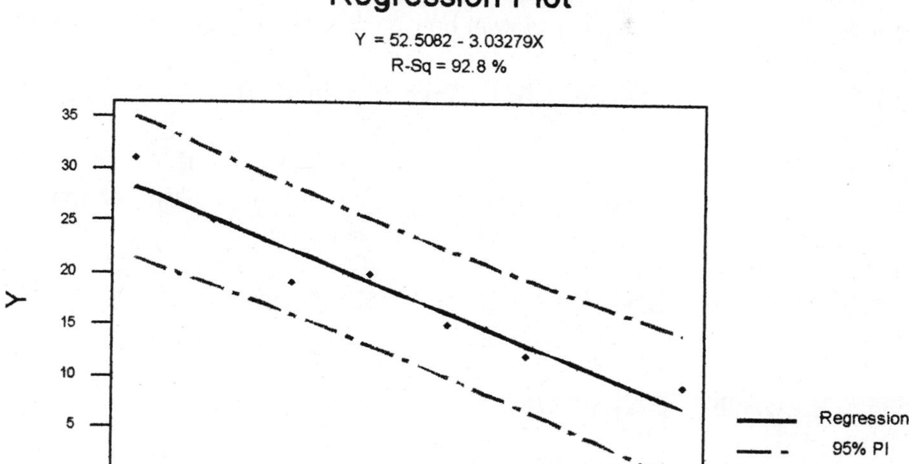

(f) Find the correlation coefficient and test it against the hypothesis that there is no correlation.
We use the menu options
➤**Stat** ➤**Basic Statistics** ➤**Correlation**

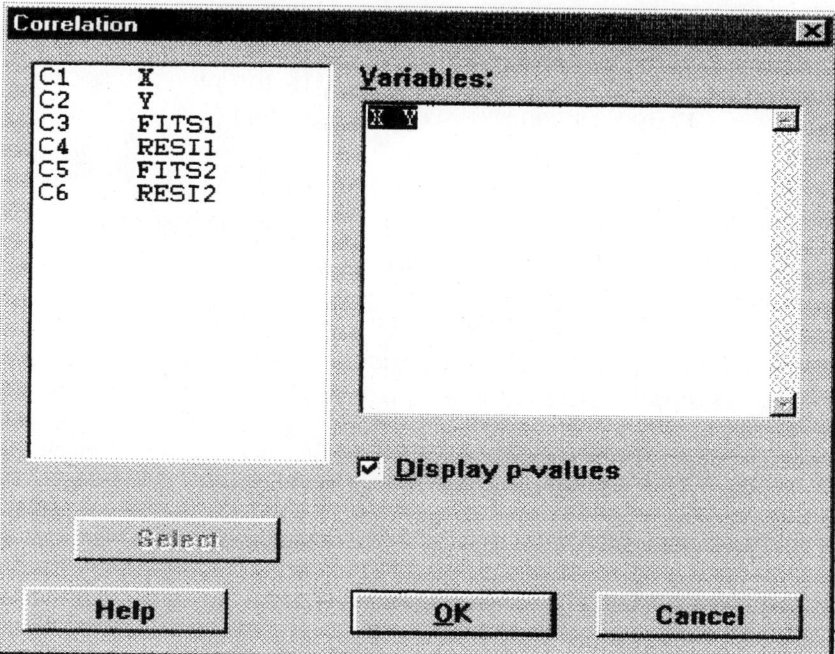

The results are

Correlations (Pearson)

```
Correlation of X and Y = -0.963, P-Value = 0.000
```

Notice r = -0.963 and the P-value is so small as to be 0. We reject the null hypothesis and conclude that there is a linear correlation between the number of clerks or duty and amount of shrinkage.

Lab Activities for Simple Regression (Two Variables)

1. Open or retrieve the worksheet TRUCK.mtp from the DATA DISK. This worksheet contains the following data, with the list price in column C1 and the best price in the column C2. The best price is the best price negotiated by a team from the magazine.

LIST PRICE VERSUS BEST PRICE FOR A NEW GMC PICKUP TRUCK

IN THE FOLLOWING DATA PAIRS (X,Y)

X = LIST PRICE (IN $1000) FOR A GMC PICKUP TRUCK

Y = BEST PRICE (IN $1000) FOR A GMC PICKUP TRUCK

SOURCE: CONSUMERS DIGEST, FEBRUARY 1994

(12.400, 11.200)	(14.300, 12.500)	(14.500, 12.700)
(14.900, 13.100)	(16.100, 14.100)	(16.900, 14.800)
(16.500, 14.400)	(15.400, 13.400)	(17.000, 14.900)
(17.900, 15.600)	(18.800, 16.400)	(20.300, 17.700)
(22.400, 19.600)	(19.400, 16.900)	(15.500, 14.000)
(16.700, 14.600)	(17.300, 15.100)	(18.400, 16.100)
(19.200, 16.800)	(17.400, 15.200)	(19.500, 17.000)
(19.700, 17.200)	(21.200, 18.600)	

a) Use MINITAB to find the least squares regression line using the best price as the response variable and list price as the explanatory variable
b) Use MINITAB to draw a scatter plot of the data.
c) What is the value of the standard error of estimate?
d) What is the value of the coefficient of determination r^2? of the correlation coefficient r?
e) Use the least squares model to predict the best price for a truck with a list price of $20,000. Note: Enter this value as 20 since X is assumed to be in thousands of dollars. Find a 95% confidence interval for the prediction.

2. Other MINITAB worksheets appropriate to use for simple linear regression are

Cricket Chirps Versus Temperature: **CRICKET.mtp**

Lab Activities for Simple Regression (Two Variables) continued

Source: *The Song of Insects* by Dr. G.W. Pierce, Harvard College Press
The chirps per second for the striped grouped cricket are stored in C1; The corresponding temperature in degrees fahrenheit is stored in C2.

Diameter of Sand Granules Versus Slope on a Natural Occurring Ocean Beach:
SAND.mtp; Source *Physical Geography* by A.M. King, Oxford press
The median diameter (MM) of granules of sand in stored in C1; The corresponding gradient of beach slope in degrees is stored in C2.

National Unemployment Rate Male Versus Female: **NUNEMPL.mtp**
Source: *Statistical Abstract of the United States*
The national unemployment rate for adult males is stored in C1; The corresponding unemployment rate for adult females for the same period of time is stored in C2.

The data in these worksheets are described in the Appendix of this *Guide*. Select these worksheets and repeat parts a-d of problem 1, using C1 as the explanatory variable and C2 as the response variable.

3. A psychologist interested in job stress is studying the possible correlation between interruptions and job stress. A clerical worker who is expected to type, answer the phone and do reception work has many interruptions. A store manager who has to help out in various departments as customer level demands also has interruptions. An accountant who is given tasks to accomplish each day and who is not expected to interact with other colleagues or customers except during specified meeting times has few interruptions. The psychologist rated a group of jobs for interruption level. She selected a random sample of 12 people holding such jobs and analyzed their stress level. The results follow with X be interruption level of the job on a scale of 1 to 20 with 20 having the most interruptions and Y the stress level on a scale of 1 to 50 with 50 the most stressed.

Person	1	2	3	4	5	6	7	8	9	10	11	12
X	9	15	12	18	20	9	5	3	17	12	17	6
Y	20	37	45	42	35	40	20	10	15	39	32	25

a) Enter the X values into C1 and the Y values into C2. Use the menu selections ➤**Stat** ➤**Basic Statistics** ➤**Display Descriptive Statistics** to on the two columns. What is the mean of the Y values? Of the X values? What are the respective standard deviations?

b) Make a scatter plot of the data using the ➤**Stat** ➤**Regression** ➤**Fitted Line** menu selection. From the diagram do you expect a positive or negative correlation?

c) Use the ➤**Stat** ➤**Basic Statistics** ➤**Correlation** menu choicesto get the value of r. Is this value consistent with your response in part b?

d) Use the ➤**Stat** ➤**Regression** ➤**Regression** menu choices with Y as the response variable and X as the explanatory variable. Use the [Option] button with

Lab Activities for Simple Regression (Two Variables) continued

predictions 5 10 15 20 to get the predicted stress level of jobs with interruption levels of 5 10 15 20. Look at the 95% P.I. intervals. Which are the longest? Why would you expect these results? Find the standard error of estimate. Is R-sq equal to the square of r as you found in part c? What is the equation of the least squares line?

e) Redo the ➤**Stat** ➤**Regression** ➤**Regression** menu option, this time using X as the response variable and Y as the explanatory variable. Is the equation different than that of part d? What about the value of the standard error of estimate (s on your output), did it change? Did R-sq change?

4. The researcher of problem 3 was able to add to her data. Another random sample of 11 people had their jobs rated for interruption level and were then evaluated for stress level.

Person	13	14	15	16	17	18	19	20	21	22	23
X	4	15	19	13	10	9	3	11	12	15	4
Y	20	35	42	37	40	23	15	32	28	38	12

Add this data to the data in problem 3, and repeat parts a through e. Compare the values of s the standard error of estimate in parts d. Did more data tend to reduce the value of s? Look at the 95% P.I. intervals. How do they compare to the corresponding ones of problem 3? Are they shorter or longer? Why would you expect this result?

MULTIPLE REGRESSION

An introduction to multiple regression is presented in Section 10.5 of *Understandable Statistics*. The **Stat ➤ Regression ➤ Regression** menu choices also does multiple regression.

Stat ➤ Regression ➤ Regression
Dialogue Box Responses
Response: Enter the column number C# of the column containing the responses (that is Y values)
Predictor: Enter the column number C# of the columns containing the explanatory variables
[Graphs]: Do not click on at this time
[Results]: Click on and select the second option, Regression equation, etc
[Options]: We will click on this option when we wish to do predictions for new variables.
[Storage]: Click on and select the fits and residual options if you wish

Example 2

Bowman Brothers is a large sporting goods store in Denver that has a giant ski sale every year during the month of October. The chief executive officer at Bowman Brothers is studying the following variables regarding the ski sale
X_1 = Total dollar receipts from October ski sale
X_2 = Total dollar amount spent advertising ski sale on local TV
X_3 = Total dollar amount spent advertising ski sale on local radio
X_4 = Total dollar amount spent advertising ski sale in Denver newspapers

Data for the past eight years is shown below (in thousands of dollars)

Year	1	2	3	4	5	6	7	8
X 1	751	768	801	832	775	718	739	780
X 2	19	23	27	32	25	18	20	24
X 3	14	17	20	24	19	9	10	19
X 4	11	15	16	18	12	5	7	14

a) Enter the data in C1-C4. Name C1 = 'X1', C2 = 'X2', C3 = 'X3' C4 = 'X4'. Use to study the data.

	N	MEAN	MEDIAN	TRMEAN	STDEV	SEMEAN
X1	8	770.5	771.5	770.5	35.8	12.6
X2	8	23.50	23.50	23.50	4.63	1.64
X3	8	16.50	18.00	16.50	5.15	1.82
X4	8	12.25	13.00	12.25	4.46	1.58

	MIN	MAX	Q1	Q3
X1	718.0	832.0	742.0	795.7
X2	18.00	32.00	19.25	26.50
X3	9.00	24.00	11.00	19.75
X4	5.00	18.00	8.00	15.75

b) Next use ➤**Stat** ➤**Basic Statistics** ➤**Correlation** menu option to see the correlation between each pair of columns of data.

	X1	X2	X3
X2	0.974		
X3	0.968	0.934	
X4	0.937	0.864	0.950

c) Use X1 as the response variable with 3 explanatory variables X2, X3, X4. Use toe [Options] button and select Predictions 21 11 8 so that you can see the predicted value of X1 for X2 = 21 X3 = 11 and X4 = 8. For this regression model, note the lease squares equation, the standard error of estimate, and the coefficient of multiple determination R-sq. Look at the P-values of the coefficients. Remember we are testing the null hypothesis H_0: $\beta_i = 0$ against the alternate hypothesis H_1:$\beta_i \neq 0$. A P-value less than α is evidence to reject H_0.

The regression equation is
X1 = 618 + 4.70 X2 + 0.65 X3 + 2.58 X4

Predictor	Coef	Stdev	t-ratio	p
Constant	617.72	14.92	41.40	0.000
X2	4.698	1.369	3.43	0.027
X3	0.652	1.979	0.33	0.758
X4	2.580	1.623	1.59	0.187

s = 5.866 R-sq = 98.5% R-sq(adj) = 97.3%

```
Analysis of Variance
SOURCE          DF          SS          MS          F          p
Regression       3      8820.3      2940.1      85.43      0.000
Error            4       137.7        34.4
Total            7      8958.0

SOURCE          DF      SEQ SS
X2               1      8497.6
X3               1       235.7
X4               1        87.0
```

Note that we will not use the results of the Analysis of variance

The predicted values are

```
     Fit  Stdev.Fit          95% C.I.            95% P.I.
  746.78       5.00   ( 732.89, 760.67)   ( 725.37, 768.19)
```

Lab Activities for Multiple Regression

Use the Section 10.5 problems 3 - 6. Each of these problems have MINITAB worksheets stored on the DATA DISK.

Section 10.5 problem #3 (Systolic Blood Pressure Data)
DATA DISK FOR MINITAB WORKSHEET NAME: **C10S5P3.mtp**
Section 10.5 problem #4 (Test Scores for General Psychology)
DATA DISK FOR MINITAB WORKSHEET NAME: **C10S5P3.mtp**
Section 10.5 problem #5 (Hollywood Movies data)
DATA DISK FOR MINITAB WORKSHEET NAME: **C10S5P5.mtp**
Section 10.5 problem #6 (All Greens Franchise Data)
DATA DISK FOR MINITAB WORKSHEET NAME: **C10S5P6.mtp**

Two additional case studies are available on the DATA DISK. The data are listed in the Appendix. For each of these studies, explore the realtionships among the variables.

MINITAB DATA DISK WORKSHEET HEALTH.mtp

THIS IS A CASE STUDY OF PUBLIC HEALTH, INCOME, AND POPULATION DENSITY FOR SMALL CITIES IN EIGHT MIDWESTERN STATES: OHIO, INDIANA, ILLINOIS, IOWA, MISSOURI, NEBRASKA, KANSAS, AND OKLAHOMA. THE DATA IS FOR A SAMPLE OF 53 SMALL CITIES IN THESE STATES.

X1 = DEATH RATE PER 1000 RESIDENTS
X2 = DOCTOR AVAILABILITY PER 100,000 RESIDENTS
X3 = HOSPITAL AVAILABILITY PER 100,000 RESIDENTS
X4 = ANNUAL PER CAPITA INCOME IN THOUSANDS OF DOLLARS
X5 = POPULATION DENSITY PEOPLE PER SQUARE MILE

MINITAB DATA DISK WORKSHEET CRIME.mtp

THIS IS A CASE STUDY OF EDUCATION, CRIME, AND POLICE FUNDING FOR SMALL CITIES IN TEN EASTERN AND SOUTH EASTERN STATES. THE STATES ARE NEW HAMPSHIRE, CONNECTICUT, RHODE ISLAND, MAINE, NEW YORK, VIRGINIA, NORTH CAROLINA, SOUTH CAROLINA, GEORGIA, AND FLORIDA. THE DATA IS FOR A SAMPLE OF 50 SMALL CITIES IN THESE STATES.

X1 = TOTAL OVERALL REPORTED CRIME RATE PER 1MILLION RESIDENTS
X2 = REPORTED VIOLENT CRIME RATE PER 100,000 RESIDENTS
X3 = ANNUAL POLICE FUNDING IN DOLLARS PER RESIDENT
X4 = PERCENT OF PEOPLE 25 YEARS AND OLDER THAT HAVE HAD 4 YEARS

OFHIGH SCHOOL
X5 = PERCENT OF 16 TO 19 YEAR-OLDS NOT IN HIGHSCHOOL AND NOT HIGH SCHOOL GRADUATES.
X6 = PERCENT OF 18 TO 24 YEAR-OLDS ENROLLED IN COLLEGE
X7 = PERCENT OF PEOPLE 25 YEARS AND OLDER WITH AT LEAST 4 YEARS OF COLLEGE

COMMAND SUMMARY

To perform simple or multiple regression

REGRESS C on K explanatory variables in C...C does regression with the first column containing the response variable, K explanatory variables in the remaining columns.

PREDICT E...E predicts the response variable for the given values of the explanatory variable(s).

RESIDUALS put into C stores the residuals in column C

WINDOWS menu selection: **Stat ➤ Regression ➤ Regression**
Use the dialog box to list the response and explanatory (prediction) variables. Mark the residuals box. In the Options dialog box list the values of the explanatory variable(s) for which you wish to make a prediction. Select the P.I. confidence interval.

BRIEF K controls the amount of output for K = 1, 2, 3 with 3 giving the most output.
This command is not available from a menu.

There are other subcommands for REGRESS. See the MINITAB Help for your release of MINITAB for a list of the subcommands and their descriptions.

To find the Pearson product moment correlation coefficient

CORRELATION for C...C calculates the correlation coefficient for all pairs of columns

WINDOWS menu selection: **Stat ➤ Basic Statistics ➤ Correlation**

To graph the scatter plot for simple regression

With **GSTD** use the **PLOT C vs C** command

WINDOWS menu selection (MINITAB SE for Windows or MINITAB 10 for Windows):
Stat ➤ Regression ➤ Fitted Line Plot

CHAPTER 11 CHI SQUARE AND F DISTRIBUTIONS

CHI SQUARE TESTS OF INDEPENDENCE

Use of the chi square distribution to test independence is discussed in Section 11.1 of *Understandable Statistics*. In such tests we use the hypotheses

H_0: The variables are independent
H_1: The variables are not independent

To use MINITAB for tests of independence, we enter the values of a contingency table row by row. The command CHISQUARE then prints a contingency table showing both the observed and expected counts. It computes the sample chi square value using the following formula in which E stands for the expected count in a cell and O stands for the observed count in that same cell. The sum is taken over all cells.

$$X = \sum \frac{(O - E)^2}{E}$$

Then MINITAB gives the number of degrees of the chi square distribution. To conclude the test use the P value of the sample chi square statistic if your version of MINITAB provides it. Otherwise compare the calculated chi square value to a table of the chi square distribution with the indicated degrees of freedom. We may use Table 8 of Appendix II of *Understandable Statistics*. If the calculated sample chi square value is larger than the value in Table 8 for a specified level of significance, we reject H_0.

Use the menu selection

Stat ➤ Tables ➤ Chisquare Test
Dialogue Box Responses
List the columns containing the data from the contingency table. Note that you mayuse up to seven columns of data. Each column must contain integer values

Example

A computer programming aptitude test has been developed for high school seniors. The test designers claim that scores on the test are independent of the type of school the student attends: rural, suburban, urban. A study involving a random sample of students from these types of institutions yielded the following contingency table. Use the CHISQUARE command to compute the sample chi square value, and to determine the degrees of freedom of the chi square distribution. Then determine if type or school and test score are independent at the $\alpha = 0.05$ level of significance.

School Type

Score	Rural	Suburban	Urban
200-299	33	65	82
300-399	45	79	95
400-500	21	47	63

To use the menu selection ➤**Stat** ➤**Tables** ➤**Chisquare Test** with C1 containing test scores for rural schools, C2 corresponding test scores for suburban schools, and C3 corresponding test scores for urban schools.

Chi-Square Test

```
Expected counts are printed below observed counts

         Rural Suburban    Urban    Total
    1       33       65       83      181
         33.75    65.11    82.15

    2       45       79       95      219
         40.83    78.77    99.40

    3       21       47       63      131
         24.42    47.12    59.46

Total       99      191      241      531

Chi-Sq =  0.016 +  0.000 +  0.009 +
          0.426 +  0.001 +  0.194 +
          0.480 +  0.000 +  0.211 = 1.338
DF = 4, P-Value = 0.855
```

Lab Activities Using CHI SQUARE TEST

Use MINITAB to produce a contingency table and compute the sample chi square value. If your version of MINITAB produces the P value of the sample chi square statistic, conclude the test using P values. Otherwise use Table 9 of *Understandable Statistics* to find the chi square value for the given α and degrees of freedom. Compare the sample chi square value to the value found in Table 8 to conclude the test.

1. We Care Auto Insurance had its staff of actuaries conduct a study to see if vehicle type and loss claim are independent. A random sample of auto claims over the 1st six months give the information in the contingency table.

Total Loss Claims per Year per Vehicle

Type of vehicle	$0-999	$1000-2999	$3000-5900	$6000+
Sports car	20	10	16	8
Truck	16	25	33	9
Family Sedan	40	68	17	7
Compact	52	73	48	12

Test the claim that car type and loss claim are independent. Use $\alpha = 0.05$.

2. An educational specialist is interested in comparing three methods of instruction:

S.L.- standard lecture with discussion
T.V.- video taped lectures with no discussion
I.M.- individualized method with reading assignments and tutoring, but no lectures.

The specialist conducted a study of these three methods to see if they are independent. A course was taught using each of the three methods and a standard final exam given at the end. Students were put into the different method sections at random. The course type and test results are shown in the contingency table.

Final Exam Score

Course Type	below 60	60-69	70-79	80-89	90-100
S.L.	10	4	70	31	25
T.V.	8	3	62	27	23
I.M.	7	2	58	25	22

Test the claim that the instruction method and final exam test scores are independent using $\alpha = 0.01$.

ANALYSIS OF VARIANCE (ANOVA)

Section 11.4 of *Understandable Statistics* introduces single factor analysis of variance (also called one-way ANOVA). We consider several populations which are each assumed to follow a normal distribution. The standard deviations of the populations are assumed to be approximately equal. ANOVA provides a method to compare several different populations to see if the means are the same. Let population 1 have mean μ_1, population 2 have mean μ_2, and so forth. The hypotheses of ANOVA are

H_0: $\mu_1 = \mu_2 = \ldots = \mu_n$
H_1: not all the means are equal

In MINITAB we use the menu se;ectopm **Stat ➤ ANOVA ➤ Oneway(Unstacked)** to perform one way ANOVA. We put the data from each population in a separate column. The different populations are called levels in the output of AOVONEWAY. An analysis of variance table is printed as well a 95% confidence interval for the mean of each level. If there are only two populations, AOVONEWAY is equivalent to using the 2-Sample Test choice with the equal variances option checked.

Stat ➤ ANOVA ➤ Oneway(Unstacked)
Dialogue Box Responses
Responses: Enter the columns containing the data.

Example

A psychologist has developed a series of tests to measure depression level. The composite scores range from 50 to 100 with 100 representing the most severe depression level. This measuring device was used in a study of treatments for depression. A random sample of 12 patients with approximately the same depression level as measured by the tests was divided into 3 different treatment groups. Then, one month after treatment was competed, the depression level of each patient was again evaluated using the series of tests. The after treatment depression levels are given

Treatment 1: 70 65 82 83 71
Treatment 2: 75 62 81
Treatment 3: 77 60 80 75

Put treatment 1 responses in column C1, treatment 2 responses in C2, treatment 3 responses in C3. Use the **Stat ➤ ANOVA ➤ Oneway(Unstacked)** menu selections

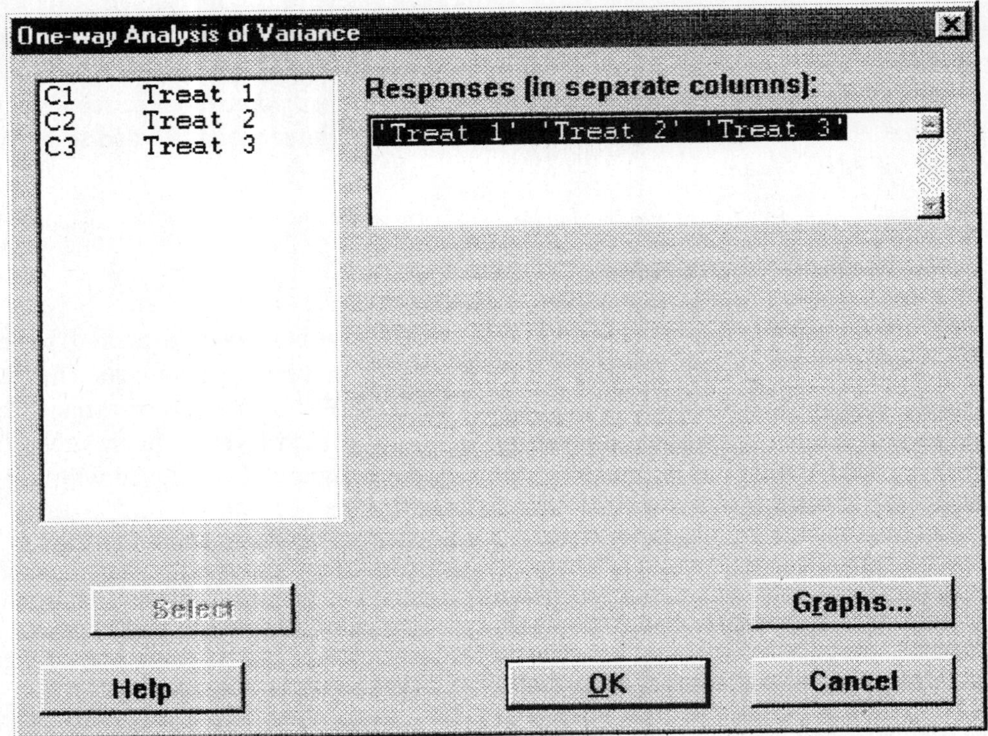

The results are

One-way Analysis of Variance

```
ANALYSIS OF VARIANCE
SOURCE      DF        SS        MS        F        p
FACTOR       2       5.4       2.7     0.04    0.965
ERROR        9     677.5      75.3
TOTAL       11     682.9
                                   INDIVIDUAL 95 PCT CI'S FOR MEAN
                                   BASED ON POOLED STDEV
LEVEL   N      MEAN      STDEV   ---+---------+---------+---------+---
 C1     5    74.200      7.918          (------------*------------)
 C2     3    72.667      9.713     (---------------*---------------)
 C3     4    73.000      8.907       (-------------*-------------)
                                 ---+---------+---------+---------+---
 POOLED STDEV =     8.676          63.0      70.0      77.0      84.0
```

Since the level of significance α = 0.05 is less than the P-value of 0.965, we no not reject H_0.

Lab Activities for Analysis of Variance ANOVA

1. A random sample of 20 overweight adults were randomly divided into 4 groups. Each group was given a different diet plan, and the weight loss for each individual after 3 months follows:

   Plan 1: 18 10 20 25 17
   Plan 2: 28 12 22 17 16
   Plan 3: 16 20 24 8 17
   Plan 4: 14 17 18 5 16

   Test the claim that the population mean weight loss is the same for the four diet plans at the 5% level of significance.

2. A psychologist is studying the time it takes rats to respond to stimuli after being given doses of different tranquilizing drugs. A random sample of 18 rats were divided into 3 groups. Each group was given a different drug. The response time to stimuli was measured (in seconds). The results follow.

   Drug A 3.1 2.5 2.2 1.5 0.7 2.4
   Drug B 4.2 2.5 1.7 3.5 1.2 3.1
   Drug C 3.3 2.6 1.7 3.9 2.8 3.5

   Test the claim that the population mean response times for the three drugs is the same at the 5% level of significance.

3. A research group is testing various chemical combinations designed to neutralize and buffer the effects of acid rain on lakes. A random sample of 18 lakes of similar size in the same region have all been affected in the same way by acid rain. The lakes are divided into four groups and each group of lakes is sprayed with a different chemical combination. An acidity index is then taken after treatment. The index ranges from 60 to 100 with 100 indicating the greatest acid rain pollution. The results follow.

   Combination I 63 55 72 81 75
   Combination II 78 56 75 73 82
   Combination III 59 72 77 60
   Combination IV 72 81 66 71

   Test the claim that the population mean acidity index after each of the four treatments is the same at the 0.01 level of significance.

COMMAND SUMMARY

CHISQUARE test on table stored in C...C produced a contingency table and computes the sample chi square value

WINDOWS menu selection: **Stat ➤ Tables ➤ Chisquare Test**
In the dialog box specify the columns which contain the chi square table.

AOVONEWAY on C...C performs a one-way analysis of variance. Each column contains data from a different population

WINDOWS menu selection: **Stat ➤ ANOVA ➤ Oneway (Unstacked)**
In the dialog box specify the columns to be included.

COMMAND REFERENCE

This appendix summarizes all the MINITAB commands used in this Guide. A complete list of commands may be found in the MINITAB Reference Manual that comes with the MINITAB software.

C denotes a column
E denotes either a column or constant
K denotes a constant
[] denotes optional parts of the command

General Information

HELP gives general information about MINITAB
WINDOWS menu: **Help**

INFO gives the status of the worksheet

STOP ends MINITAB session
WINDOWS menu selection: **File ➤ Exit**

To Enter Data

READ C...C puts data into designated columns
READ 'filename' C...C reads data from file into columns
SET C puts data into single designated column
SET 'filename' C reads data from file into column
END signals end of data
NAME C = 'name' names column C
WINDOWS menu selection: You can enter data in rows or columns and name the column in the DATA window. To access the data window select **Window ➤ Data**

RETRIEVE 'filename' retrieves worksheet
WINDOWS menu selection: **File ➤ Retrieve**

To Edit Data

LET C(K) = K changes the value in row K of column C
INSERT K K C C inserts data between rows K and K into columns C to C
DELETE K K C C deletes data between rows K and K from columns C to C

WINDOWS menu selection: You can edit data in rows or columns in the DATA window. To access the data window select **Window ➤ Data**

COPY C into C copies column C into column C
USE rows K...K subcommand to copy designated rows
OMIT rows K...K subcommand to omit designated rows

WINDOWS menu selection: **Manip ➤ Copy Columns**

ERASE E...E erases designated columns or constants
WINDOWS menu selection: **Manip ➤ Erase Variables**

<u>To Output Data</u>

PRINT E...E prints designated columns or constant
WINDOWS menu selection: **File ➤ Display Data**

SAVE 'filename' saves current worksheet
PORTABLE subcommand to make worksheet portable
WINDOWS menu selection: **File ➤ Save Worksheet**
WINDOWS menu selection: **File ➤ Save Worksheet As...** you may select portable

WRITE 'filename' C...C saves data in ASCII file
WINDOWS menu selection: **File ➤ Other Files ➤ Export ASCII Data**

<u>Miscellaneous</u>

PAPER prints session
NOPAPER stops printing session
WINDOWS menu selection: **File ➤ Print Window**

OUTFILE = 'filename' saves session in ASCII file
NOOUTFILE ends OUTFILE
WINDOWS menu selection: **File ➤ Other Files ➤ Start/Stop Recording**

<u>To do arithmetic</u>

LET E = expression evaluates the expression and stores the result in E where E may be a column or a constant

** raises to a power
* multiplication
/ division
+ addition
- subtraction

SQRT(E) takes the square root
ROUND(E) rounds numbers to the nearest integer
There are other arithmetic operations possible.
WINDOW menu selection: **Calc ➤ Calculator**

<u>To generate a random sample</u>

RANDOM K into C...C selects a random sample from the distribution described in the subcommand
INTEGER K to K distribution of integers from K to K
BERNOULLI P = K

BINOMIAL N = K, P = K
CHISQUARE degrees of freedom = K
DISCRETE values in C probabilities in C
F df numerator = K, df denominator = K
NORMAL [mean = K [standard deviation = K]]
POISSON mean = K
T degrees of freedom = K
UNIFORM continuous distribution on [K to K]

WINDOWS menu selection: **Calc ➤ Random data ➤ Select Distribution**

SAMPLE K rows from C...C and put results in C...C takes a random sample of rows without replacement
REPLACE causes the sample to be taken with replacement

To organize data

SORT C...C put in C...C sorts the data in the first column and carries the other columns along
DESCENDING C...C subcommand to sort in descending order
WINDOWS menu selection: **Manip ➤ Sort**

TALLY data in C...C tallies data in columns. The data must be integer valued
COUNTS
PERCENTS
CUMCOUNTS
CUMPERCENTS
ALL gives all four values
WINDOWS menu selection: **Stats ➤ Tables ➤ Tally**

HISTOGRAM C...C prints a separate histogram for data in each of the listed columns
START with midpoint = K [end with midpoint = K]
INCREMENT = K specifies distance between midpoints
WINDOWS menu selection: (for professional graphics) **Graph ➤ Histogram (options for cutpoints)**
WINDOWS menu selection: (for character graphics) **Graph ➤ Character Graphs ➤ Histogram**

STEM-AND-LEAF display of C...C makes separate stem-and-leaf displays of data in each of the listed columns
INCREMENT = K sets the distance between two display lines
TRIM outliers lists extreme data on special lines
WINDOWS menu selection: **Graph ➤ Character Graphs ➤ Stem-and-Leaf**

BOXPLOT C makes a box-and-whisker plot of data in column C
START = K [end = k]
INCREMENT = K

WINDOWS menu selection: (character graphics) **Graph➤ Character Graphs ➤ Box**
WINDOWS menu selection: (professional graphics) **Graph ➤ Boxplot**

To summarize data by column

DESCRIBE C...C prints descriptive statistics
WINDOWS menu selection: **Stat ➤ Descriptive Statistics**

COUNT C [put into K] counts the values
N C [put into K] counts the non-missing values
NMIS C [put into K] counts the missing values
SUM C [put into K] sums the values
MEAN C [put into K] gives arithmetic mean of values
STDEV C [put into K] gives sample standard deviation
MEDIAN C [put into K] gives the median of the values
MINIMUM C [put into K] gives the minimum of the values
MAXIMUM C [put into K] gives the maximum of the values
SSQ C [put into K] gives the sum of squares of values

To summarize data by row

RCOUNT E...E put into C
RN E...E put into C
RNMIS E...E put into C
RSUM E...E put into C
RMEAN E...E put into C
RSTDEV E...E put into C
RMEDIAN E...E put into C
RMIN E...E put into C
RMAX E...E put into C
RSSQ E...E put into C

To find probabilities

PDF for values in E [put into E] calculates probabilities for the specified values of a discrete distribution and calculates the probability density function for a continuous distribution.

CDF for values in E...E [put into E...E] gives the cumulative distribution. For any value X CDF X gives the probability that a random variable with the specified distribution has a value less than or equal to X

INVCDF for values in E [put into E] gives the inverse of the CDF.

Each of these commands apply the following distributions (as well as some others). If no subcommand is used, the default distribution is the standard normal.

BINOMIAL n = K p = K
POISSON μ **= K** (note that for the Poisson distribution $\mu = \lambda$)
INTEGER a = K b = K
DISCRETE values in C, probabilities in C
NORMAL μ **= K** σ **= K**
UNIFORM a = K b = K
T d.f. = K
F d.f. numerator = K d.f. denominator = K
CHISQUARE d.f. = K

WINDOWS menu selection: **Calc ➤ Probability Distribution ➤ Select distribution**
In the dialogue box, select **Probability for PDF; Cumulative probability for CDF; Inverse cumulative for INV;** Enter the required information such as **E, n, p, or** μ**, d.f.** and so forth.

Graphing commands

Character Graphics Commands

Note: In some versions of Minitab, you must use the command **GSTD** before you use the following graphics commands.

PLOT C versus C prints a scatter plot with the first column on the vertical axis and the second on the horizontal axis. The following subcommands can be used with PLOT

TITLE = 'text' gives a title above the graph
FOOTNOTE = 'text' places a line of text below the graph
XLABEL = 'text' labels the x-axis
YLABEL = 'text' labels the y-axis
SYMBOL = 'symbol' selects the symbol for the points on the graph.
The default is *
XINCREMENT = K is distance between tick marks on x-axis
XSTART = K [end = k] specifies the first tick mark and optionally the last one
YINCREMENT = K is distance between tick marks on y-axis
YSTART = K [end = K] specifies the first tick mark and optionally the last one

WINDOWS menu selection: **Graph ➤ Character Graphs ➤ Scatter Plot**
Titles, labels, and footnotes are in the **Annotate...** option
Increment and start are in the **Scale** option.

Professional Graphics

Note: In some versions of Minitab, you must use the command **GPRO** before you use the following graphics commands.

Plot C * C prints a scatter plot with the first column on the vertical axis and the second on the horizontal axis. Note that the columns must be separated by an asterisk *.

Connect connects the points with a line

Other subcommands may be used to title the graph and let the tick marks on the axes. See your MINITAB software manual for details.

WINDOWS menu selection: **Graph ➤ Plot**
Use the dialog boxes to title the graph, label the axes, set the tick marks, and so forth. See your MINITAB software manual for details.

<u>Control Charts</u>

Character Graphics Commands

Note: In some versions of Minitab, you must use the command **GSTD** before you use the following graphics commands.

CHART C...C produces a control chart under the assumption that the data come from a normal distribution with mean and standard deviation specified by the subcommands

MU = K gives the mean of the normal distribution
SIGMA = K gives the standard deviation.

WINDOWS menu selection: none for character graphics. Use the commands in the session window.

Professional Graphics

Note: This may be the default mode for versions of MINITAB supporting professional graphics. If necessary, use the command **GPRO** before using the following commands.

CHART C...C produces a control chart under the assumption that the data come from a normal distribution with mean and standard deviation specified by the subcommands

MU K gives the mean of the normal distribution
SIGMA K gives the standard deviation.

WINDOWS menu selection: **Stat ➤ Control Chart ➤ Individual**
Enter choices for MU and SIGMA in the dialog box

<u>To generate confidence intervals</u>

ZINTERVAL [K% confidence] σ = K on C...C generates a confidence interval for μ using the normal distribution. You must enter a value for σ, either actual or estimated. A separate interval is given for data in each column. If K is not specified, a 95% confidence interval will be given.

WINDOWS menu selection: **Stat➤Basic Statistics➤1 sample z**
In the dialog box select confidence interval and enter the confidence level

TINTERVAL [K% confidence] for C...C generates a confidence interval for μ using the Student's t distribution. It automatically computes stdev s from the data as well as the number of degrees of freedom. If K is not specified, a 95% confidence interval is given.

<u>To test a single mean</u>

ZTEST [μ = K] σ = K, for C...C performs a z-test on the data in each column. If you do not specify μ, it is assumed to be 0. You need to supply a value for σ (either actual, or estimated by the sample standard deviation s of a column in the case of large samples). If the ALTERNATIVE subcommand is not used, a two tail test is conducted.

WINDOWS menu selection: **Stat ➤ Basic Statistics ➤ 1 sample z**
In dialog box select alternate hypothesis, specify the mean for H_0, specify the standard deviation

TTEST [μ = K] on C...C performs a separate t-test on the data of each column. If you do not specify μ, it is assumed to be 0. The computer evaluates s, the sample standard deviation for each column, and uses the computed s value to conduct the test. If the ALTERNATIVE subcommand is not used, a two tail test is conducted.

WINDOWS menu selection: **Stat ➤ Basic Statistics ➤ 1 sample t**
In dialog box select alternate hypothesis, specify the mean for H_0

ALTERNATIVE = K is the subcommand required to conduct a one tail test. If K = -1, then a left tail test is done. If K = 1, then a right tail test is done.

To test a difference of means (independent samples)

TWOSAMPLE [K% confidence] for C C does a two (independent) sample t test and (optionally confidence interval) for data in the two columns listed. The first data set is put into the first column, and the second data set into the second column. Unless the ALTERNATIVE subcommand is used, the alternate hypothesis is assumed to be $H_1 : \mu_1 \neq \mu_2$. Samples are assumed to be independent.

ALTERNATIVE = K is the subcommand to change the alternate hypothesis to a left tail test with K = -1 or right tail test with K = 1.

POOLED is the subcommand to be used only when the two samples come from populations with equal standard deviations.

WINDOWS menu selection: **Stat ➤ Basic Statistics ➤ 2 sample t**
In dialog box select alternate hypothesis, specify the mean for H_0, for small samples select equal variances.

To perform simple or multiple regression

REGRESS C on K explanatory variables in C...C does regression with the first column containing the response variable, K explanatory variables in the remaining columns.

PREDICT E...E predicts the response variable for the given values of the explanatory variable(s).

RESIDUALS put into C stores the residuals in column C

WINDOWS menu selection: **Stat ➤ Regression ➤ Regression**
Use the dialog box to list the response and explanatory (prediction) variables. Mark the residuals box. In the Options dialog box list the values of the explanatory variable(s) for which you wish to make a prediction. Select the P.I. confidence interval.

BRIEF K controls the amount of output for K = 1, 2, 3 with 3 giving the most output. This command is not available from a menu.

There are other subcommands for REGRESS. See the MINITAB reference manual for your release of MINITAB for a list of the subcommands and their descriptions.

To find the Pearson product moment correlation coefficient

CORRELATION for C...C calculates the correlation coefficient for all pairs of columns

WINDOWS menu selection: **Stat ➤ Basic Statistics ➤ Correlation**

To graph the scatter plot for simple regression

With **GSTD** use the **PLOT C vs C** command

WINDOWS menu selection (MINITAB SE for Windows or MINITAB 10 for Windows): **Stat ➤ Regression ➤ Fitted Line Plot**

To perform chi square tests and ANOVA

CHISQUARE test on table stored in C...C produced a contingency table and computes the sample chi square value

WINDOWS menu selection: **Stat ➤ Tables ➤ Chisquare Test**
In the dialog box specify the columns which contain the chi square table.

AOVONEWAY on C...C performs a one-way analysis of variance. Each column contains data from a different population

WINDOWS menu selection: **Stat ➤ ANOVA ➤ Oneway (Unstacked)**
In the dialog box specify the columns to be included.

Nonparametric commands

MANN-WHITNEY [confidence = K] on CC does a two-sample rank sum test for the difference of two population means. Data from each population is in each separate column. The test is a two tail test unless ALTERNATE subcommand is used.

WINDOWS menu selection: **Stat ➤ Nonparametrics ➤ Mann-Whitney**

APPENDIX

CLASSROOM DEMONSTRATION DATA FILES FOR *COMPUTERSTAT VERSION 5*

WITH

MINITAB WORKSHEET NAMES FOR DATA ON THE DATA DISK

AND

TI-83 FILES FOR DATA ON THE DATA DISK

Appendix

DATA FILES IN CLASS DEMONSTRATIONS FOR COMPUTERSTAT WITH WORKSHEET NAMES OF MINITAB FILES ON DATA DISK

Disney Stock Volume

THE FOLLOWING DATA REPRESENTS THE NUMBER OF SHARES OF DISNEY STOCK (IN HUNDREDS OF SHARES) SOLD FOR A RANDOM SAMPLE OF 60 TRADING DAYS IN 1993 AND 1994.

SOURCE: DOW-JONES INFORMATION RETRIEVAL SERVICE

12584	9441	18960	21480	10766	13059	8589	4965
4803	7240	10906	8561	6389	14372	18149	6309
13051	12754	10860	9574	19110	29585	21122	14522
17330	18119	10902	29158	16065	10376	10999	17950
15418	12618	16561	8022	9567	9045	8172	13708
11259	10518	9301	5197	11259	10518	9301	5197
6758	7304	7628	14265	13054	15336	14682	27804
16022	24009	32613	19111				

DATA DISK FOR MINITAB WORKSHEET NAME: DISN.mtp
TI-83 Name: DISN.TXT

Weights of Pro Football Players

THE FOLLOWING DATA REPRESENTS WEIGHTS IN POUNDS OF 50 RANDOMLY SELECTED PRO FOOTBALL LINE BACKERS.

SOURCE: THE SPORTS ENCYCLOPEDIA PRO FOOTBALL 1960-1992

225	230	235	238	232	227	244	222
250	226	242	253	251	225	229	247
239	223	233	222	243	237	230	240
255	230	245	240	235	252	245	231
235	234	248	242	238	240	240	240
235	244	247	250	236	246	243	255
241	245						

DATA DISK FOR MINITAB WORKSHEET NAME: WEIGHTS.mtp
TI-83 Name: WEIGHTS.TXT

Heights of Pro Basketball Players

THE FOLLOWING DATA REPRESENTS HEIGHTS IN FEET OF 65 RANDOMLY SELECTED PRO BASKETBALL PLAYERS.

SOURCE: ALL-TIME PLAYER DIRECTORY , THE OFFICIAL NBA ENCYCLOPEDIA

6.50	6.25	6.33	6.50	6.42	6.67	6.83	6.82
6.17	7.00	5.67	6.50	6.75	6.54	6.42	6.58
6.00	6.75	7.00	6.58	6.29	7.00	6.92	6.42
5.92	6.08	7.00	6.17	6.92	7.00	5.92	6.42
6.00	6.25	6.75	6.17	6.75	6.58	6.58	6.46
5.92	6.58	6.13	6.50	6.58	6.63	6.75	6.25
6.67	6.17	6.17	6.25	6.00	6.75	6.17	6.83
6.00	6.42	6.92	6.50	6.33	6.92	6.67	6.33
6.08							

DATA DISK FOR MINITAB WORKSHEET NAME: HEIGHTS.mtp
TI-83 Name: HEIGHTS.TXT

Miles per Gallon Gasoline Consumption

THE FOLLOWING DATA REPRESENTS MILES PER GALLON GASOLINE CONSUMPTION (HIGHWAY) FOR A RANDOM SAMPLE OF 55 MAKES AND MODELS OF PASSENGER CARS.

SOURCE: ENVIRONMENTAL PROTECTION AGENCY

30	27	22	25	24	25	24	15
35	35	33	52	49	10	27	18
20	23	24	25	30	24	24	24
18	20	25	27	24	32	29	27
24	27	26	25	24	28	33	30
13	13	21	28	37	35	32	33
29	31	28	28	25	29	31	

DATA DISK FOR MINITAB WORKSHEET NAME: MPGAL.mtp
TI-83 Name: MPGAL.TXT

Fasting Glucose Blood Tests

THE FOLLOWING DATA REPRESENTS GLUCOSE BLOOD LEVEL (mg/100ml) AFTER A 12 HOUR FAST FOR A RANDOM SAMPLE OF 70 WOMEN.
SOURCE: AMERICAN J. CLIN. NUTR. VOL. 19, 345-351

45	66	83	71	76	64	59	59
76	82	80	81	85	77	82	90
87	72	79	69	83	71	87	69
81	76	96	83	67	94	101	94
89	94	73	99	93	85	83	80
78	80	85	83	84	74	81	70
65	89	70	80	84	77	65	46
80	70	75	45	101	71	109	73
73	80	72	81	63	74		

DATA DISK FOR MINITAB WORKSHEET NAME: GLUCOS.mtp
TI-83 name: GLUCOS.TXT

Number of Children in Rural Canadian Families

THE FOLLOWING DATA REPRESENTS THE NUMBER OF CHILDREN IN A RANDOM SAMPLE OF 50 RURAL CANADIAN FAMILIES.

SOURCE: AMERICAN JOURNAL OF SOCIOLOGY VOL. 53, 470-480

11	13	4	14	10	2	5	0
0	3	9	2	5	2	3	3
3	4	7	1	9	4	3	3
2	6	0	2	6	5	9	5
4	3	2	5	2	2	3	5
14	7	6	6	2	5	3	4
6	1						

DATA DISK FOR MINITAB WORKSHEET NAME: CHILD.mtp
TI-83 name: CHILD.TXT

YIELD OF WHEAT AT ROTHAMSTED EXPERIMENT STATION, ENGLAND

THE FOLLOWING DATA REPRESENT ANNUAL YIELD OF WHEAT IN TONNES (ONE TON = 1.016 TONNE) FOR AN EXPERIMENTAL PLOT OF LAND AT ROTHAMSTED EXPERIMENT STATION U.K. OVER A PERIOD OF THIRTY CONSECUTIVE YEARS.

SOURCE: ROTHAMSTED EXPERIMENT STATION U.K.

WE WILL USE THE FOLLOWING TARGET PRODUCTION VALUES:
TARGET MU = 2.6 TONNES
TARGET SIGMA = 0.40 TONNES

1.730	1.660	1.360	1.190	2.660	2.140	2.250	2.250	2.360	2.820
2.610	2.510	2.610	2.750	3.490	3.220	2.370	2.520	3.430	3.470
3.200	2.720	3.020	3.030	2.360	2.830	2.760	2.070	1.630	3.020

DATA DISK FOR MINITAB WORKSHEET NAME: WHEAT.mtp
TI-83 name WHEAT.TXT

PepsiCo STOCK CLOSING PRICES

THE FOLLOWING DATA REPRESENT 25 WEEKLY CLOSING PRICES IN DOLLARS PER SHARE OF PepsiCo STOCK FOR MAY TO NOV. 1993

SOURCE: DOW-JONES INFORMATION RETRIEVAL SERVICE

THE LONG TERM ESTIMATES FOR WEEKLY CLOSINGS ARE
TARGET MU = 37 DOLLARS PER SHARE
TARGET SIGMA = 1.75 DOLLARS PER SHARE

37.000	36.500	36.250	35.250	35.625	36.500	37.000	36.125	35.125	37.250
37.125	36.750	38.000	38.875	38.750	39.500	39.875	41.500	40.750	39.250
39.000	40.500	39.500	40.500	37.875					

DATA DISK FOR MINITAB WORKSHEET NAME: PEPCL.mtp
TI-83 name: PEPCL.TXT

PepsiCo STOCK VOLUME OF SALES

THE FOLLOWING DATA REPRESENT VOLUME OF SALES (IN HUNDREDS OF THOUSANDS OF SHARES) OF PepsiCo STOCK FOR THE SAME PERIOD AS DEMONSTRATION #2 (MAY TO NOV. 1993)

SOURCE: DOW-JONES INFORMATION RETRIEVAL SERVICE

FOR THE LONG TERM MU AND SIGMA WE USE
TARGET MU = 15
TARGET SIGMA = 4.5

19.000	29.630	21.600	14.870	16.620	12.860	12.250	20.870	23.090	21.710
11.140	5.520	9.480	21.100	15.640	10.790	13.370	11.640	7.690	9.820
8.240	12.110	7.470	12.670	12.330					

DATA DISK FOR MINITAB WORKSHEET NAME: PEPVOL.mtp
TI-83 name: PEPVOL.TXT

FUTURES QUOTES FOR THE PRICE OF COFFEE BEANS

THE FOLLOWING DATA REPRESENT FUTURES OPTIONS QUOTES FOR THE PRICE OF COFFEE BEANS (DOLLARS PER POUND) FOR 20 CONSECUTIVE BUSINESS DAYS IN JULY 1994

SOURCE: DOW-JONES INFORMATION RETRIEVAL SERVICE

WE USE THE FOLLOWING ESTIMATED TARGET VALUES FOR PRICING
TARGET MU = $2.15
TARGET SIGMA = $0.12

2.300	2.360	2.270	2.180	2.150	2.180	2.120	2.090	2.150	2.200
2.170	2.160	2.100	2.040	1.950	1.860	1.910	1.880	1.940	1.990

DATA DISK FOR MINITAB WORKSHEET NAME: COFFEE.mtp
TI-83 name: COFFEE.TXT

INCIDENCE OF MELANOMA TUMORS

THE FOLLOWING DATA REPRESENT NUMBER OF CASES OF MELANOMA SKIN CANCER (PER 100,000 POPULATION) IN CONNECTICUT FOR EACH OF THE YEARS 1953 TO 1972.

SOURCE: INST. J. CANCER , VOL. 25, 95-104

WE WILL USE THE FOLLOWING LONG TERM VALUES (MU AND SIGMA)
TARGET MU = 3
TARGET SIGMA = 0.9

2.400 2.200 2.900 2.500 2.600 3.200 3.800 4.200 3.900 3.700
3.300 3.700 3.900 4.100 3.800 4.700 4.400 4.800 4.800 4.800

DATA DISK FOR MINITAB WORKSHEET NAME: TUMOR.mtp
TI-83 name: TUMOR.TXT

PERCENT CHANGE IN CONSUMER PRICE INDEX

THE FOLLOWING DATA REPRESENT ANNUAL PERCENT CHANGE IN CONSUMER PRICE INDEX FOR EACH YEAR 1963 TO 1992.

SOURCE: STATISTICAL ABSTRACT OF THE UNITED STATES

SUPPOSE AN ECONOMIST RECOMMENDS THE FOLLOWING LONG TERM TARGET VALUES FOR MU AND SIGMA
TARGET MU = 4.0%
TARGET SIGMA = 1.0%

1.300 1.300 1.600 2.900 3.100 4.200 5.500 5.700 4.400 3.200
6.200 11.000 9.100 5.800 6.500 7.600 11.300 13.500 10.300 6.200
3.200 4.300 3.600 1.900 3.600 4.100 4.800 5.400 4.200 3.000

DATA DISK FOR MINITAB WORKSHEET NAME: CPI.mtp
TI-83 name: CPI.TXT

Number of Pups in Wolf Den

THE FOLLOWING DATA REPRESENT THE NUMBER OF WOLF PUPS PER DEN FROM A SAMPLE OF 16 WOLF DENS.

SOURCE: THE WOLF IN THE SOUTHWEST: THE MAKING OF AN ENDANGERED SPECIES, D.E. BROWN, UNIV. AZ. PRESS

DATA

5.00	8.00	7.00	5.00	3.00	4.00	3.00	9.00
5.00	8.00	5.00	6.00	5.00	6.00	4.00	7.00

DATA DISK FOR MINITAB WORKSHEET NAME: WOLF.mtp
TI-83 name: WOLF.TXT

Glucose Blood Level for A Specific Woman

THE FOLLOWING DATA REPRESENT GLUCOSE BLOOD LEVEL (mg/100ml) AFTER A 12 HOUR FAST FOR SIX TESTS GIVEN TO AN ADULT WOMAN.

SOURCE: AMERICAN J. CLIN. NUTR. VOL. 19, 345-351

DATA

83.00	83.00	86.00	86.00	78.00	88.00

DATA DISK FOR MINITAB WORKSHEET NAME: SGLUC.MTP
TI-83 name: SGLUC.TXT

Heights of Football Players Versus Heights of Basketball Players

THE FOLLOWING DATA REPRESENT HEIGHTS IN FEET OF 45 RANDOMLY SELECTED PRO FOOTBALL PLAYERS, AND 40 RANDOMLY SELECTED PRO BASKETBALL PLAYERS.

SOURCE: SPORTS ENCYCLOPEDIA PRO FOOTBALL, AND THE OFFICIAL NBA BASKETBALL ENCYCLOPEDIA

X1 = HEIGHTS (FT.) OF PRO FOOTBALL PLAYERS ; MU1 = POPULATION MEAN FOR X1

6.33	6.50	6.50	6.25	6.50	6.33	6.25	6.17	6.42	6.33
6.42	6.58	6.08	6.58	6.50	6.42	6.25	6.67	5.91	6.00
5.83	6.00	5.83	5.08	6.75	5.83	6.17	5.75	6.00	5.75
6.50	5.83	5.91	5.67	6.00	6.08	6.17	6.58	6.50	6.25
6.33	5.25	6.67	6.50	5.83					

X2 = HEIGHTS (FT.) OF PRO BASKETBALL PLAYERS ; MU2 = POPULATION MEAN FOR X2

6.08	6.58	6.25	6.58	6.25	5.92	7.00	6.41	6.75	6.25
6.00	6.92	6.83	6.58	6.41	6.67	6.67	5.75	6.25	6.25
6.50	6.00	6.92	6.25	6.42	6.58	6.58	6.08	6.75	6.50
6.83	6.08	6.92	6.00	6.33	6.50	6.58	6.83	6.50	6.58

DATA DISK FOR MINITAB WORKSHEET NAME: FHVBH.mtp
TI-83 name: FHVBH.TXT

Weight of Football Players Versus Weight of Basketball Players

THE FOLLOWING DATA REPRESENT WEIGHTS IN POUNDS OF 21 RANDOMLY SELECTED PRO FOOTBALL PLAYERS, AND 19 RANDOMLY SELECTED PRO BASKETBALL PLAYERS.

SOURCE: SPORTS ENCYCLOPEDIA PRO FOOTBALL, AND THE OFFICIAL NBA BASKETBALL ENCYCLOPEDIA

X1 = WEIGHTS (LB.) OF PRO FOOTBALL PLAYERS ; MU1 = POPULATION MEAN FOR X1

245 262 255 251 244 276 240 265 257 252 282
256 250 264 270 275 245 275 253 265 270

X2 = WEIGHTS (LB.) OF PRO BASKETBALL PLAYERS ; MU2 = POPULATION MEAN FOR X2

205 200 220 210 191 215 221 216 228 207
225 208 195 191 207 196 181 193 201

DATA DISK FOR MINITAB WORKSHEET NAME: FWVBW.mtp
TI-83 name: FWVBW.TXT

Petal Length for Iris Virginica Versus Petal Length for Iris Setosa

THE FOLLOWING DATA REPRESENT PETAL LENGTH (CM.) FOR A RANDOM SAMPLE OF 35 IRIS VIRGINICA AND A RANDOM SAMPLE OF 38 IRIS SETOSA

SOURCE: ANDERSON, E., BULL. AMER. IRIS SOC.

X1 = PETAL LENGTH (C.M.) IRIS VIRGINICA ; MU1 = POPULATION MEAN FOR X1

5.1 5.8 6.3 6.1 5.1 5.5 5.3 5.5 6.9 5.0 4.9 6.0 4.8 6.1 5.6 5.1
5.6 4.8 5.4 5.1 5.1 5.9 5.2 5.7 5.4 4.5 6.1 5.3 5.5 6.7 5.7 4.9
4.8 5.8 5.1

X2 = PETAL LENGTH (C.M.) IRIS SETOSA ; MU2 = POPULATION MEAN FOR X2

1.5 1.7 1.4 1.5 1.5 1.6 1.4 1.1 1.2 1.4 1.7 1.0 1.7 1.9 1.6 1.4
1.5 1.4 1.2 1.3 1.5 1.3 1.6 1.9 1.4 1.6 1.5 1.4 1.6 1.2 1.9 1.5
1.6 1.4 1.3 1.7 1.5 1.7

DATA DISK FOR MINITAB WORKSHEET NAME: PETAL.mtp
TI-83 name: PETAL.TXT

SEPAL WIDTH OF IRIS VERSICOLOR VERSUS IRIS VIRGINICA

THE FOLLOWING DATA REPRESENT SEPAL WIDTH (CM.) FOR A RANDOM SAMPLE OF 40 IRIS VERSICOLOR AND A RANDOM SAMPLE OF 42 IRIS VIRGINICA

SOURCE: ANDERSON, E., BULL. AMER. IRIS SOC.

WE USE ALPHA = 0.05 ; H0:MU1=MU1 ; AND H1:MU1<>MU2
X1 = SEPAL WIDTH (C.M.) IRIS VERSICOLOR MU1 = POPULATION MEAN FOR X1

3.2 3.2 3.1 2.3 2.8 2.8 3.3 2.4 2.9 2.7 2.0 3.0 2.2 2.9 2.9 3.1
3.0 2.7 2.2 2.5 3.2 2.8 2.5 2.8 2.9 3.0 2.8 3.0 2.9 2.6 2.4 2.4
2.7 2.7 3.0 3.4 3.1 2.3 3.0 2.5

X2 = SEPAL WIDTH (C.M.) IRIS VIRGINICA MU2 = POPULATION MEAN FOR X2

3.3 2.7 3.0 2.9 3.0 3.0 2.5 2.9 2.5 3.6 3.2 2.7 3.0 2.5 2.8 3.2
3.0 3.8 2.6 2.2 3.2 2.8 2.8 2.7 3.3 3.2 2.8 3.0 2.8 3.0 2.8 3.8
2.8 2.8 2.6 3.0 3.4 3.1 3.0 3.1 3.1 3.1

DATA DISK FOR MINITAB WORKSHEET NAME: SEPALW.mtp
TI-83 name: SEPAL.TXT

Number of Cases of Rabies in Region 1 Versus Region 2

THE FOLLOWING DATA REPRESENT NUMBER OF CASES OF RED FOX RABIES FOR A RANDOM SAMPLE OF 16 AREAS IN EACH OF TWO DIFFERENT REGIONS OF SOUTHERN GERMANY.

SOURCE: SAYERS, B., MEDICAL INFORMATICS VOL. 2, 11-34

X1 = NUMBER CASES IN REGION 1 ; MU1 = POPULATION MEAN FOR X1

10 2 2 5 3 4 3 3 4 0 2 6 4 8 7 4

X2 = NUMBER CASES IN REGION 2 ; MU2 = POPULATION MEAN FOR X2

1 1 2 1 3 9 2 2 4 5 4 2 2 0 0 2

DATA DISK FOR MINITAB WORKSHEET NAME: RABIES.mtp
TI-83 name: RABIES.TXT

Average Faculty Salary for Males Versus Females (Paired Data)

IN THE FOLLOWING DATA PAIRS (A,B)
A = AVERAGE SALARY FOR MALES ($1000/YR)
B = AVERAGE SALARY FOR FEMALES ($1000/YR)
FOR ASSISTANT PROFESSORS AT THE SAME COLLEGE OR UNIVERSITY.
A RANDOM SAMPLE OF 22 U.S. COLLEGES AND UNIVERSITIES
WAS USED. WE LET MU = POPULATION MEAN OF D = B-A VALUES

SOURCE: ACADEME, BULLETIN OF THE AMERICAN ASSOCIATION
OF UNIVERSITY PROFESSORS

WE USE N = 22 WITH ALPHA = 0.05
NULL HYPOTHESIS H0: MU = 0
ALTERNATE HYPOTHESIS H1: MU <> 0

(34.5,33.9) (30.5,31.2) (35.1,35.0) (35.7,34.2) (31.5,32.4)
(34.4,34.1) (32.1,32.7) (30.7,29.9) (33.7,31.2) (35.3,35.5)
(30.7,30.2) (34.2,34.8) (39.6,38.7) (30.5,30.0) (33.8,33.8)
(31.7,32.4) (32.8,31.7) (38.5,38.9) (40.5,41.2) (25.3,25.5)
(28.6,28.0) (35.8,35.1)

DATA DISK FOR MINITAB WORKSHEET NAME: FSALARY.mtp
TI-83 name: FSALARY.TXT

Unemployment for College Graduates Versus High School Only (Paired Data)

IN THE FOLLOWING DATA PAIRS (A,B)
A = PERCENT UNEMPLOYMENT FOR COLLEGE GRADUATES
B = PERCENT UNEMPLOYMENT FOR HIGH SCHOOL ONLY
THE DATA IS PAIRED BY YEAR STARTING WITH 1991 BACK
FOR A TOTAL OF 12 YEARS. LET D=B-A AND MU=POPULATION
MEAN OF ALL D VALUES.

SOURCE: STATISTICAL ABSTRACT OF THE UNITED STATES

WE WILL USE N = 12 WITH ALPHA = 0.001
NULL HYPOTHESIS H0: MU = 0
ALTERNATE HYPOTHESIS H1: MU > 0

(2.8, 5.9) (2.2, 4.9) (2.2, 4.8) (1.7, 5.4) (2.3, 6.3)
(2.3, 6.9) (2.4, 6.9) (2.7, 7.2) (3.5,10.0) (3.0, 8.5)
(1.9, 5.1) (2.5, 6.9)

DATA DISK FOR MINITAB WORKSHEET NAME: UNEMPL.mtp
TI-83 name: UNEMPL.TXT

Number of Navajo Hogans Versus Modern Houses (Paired Data)

IN THE FOLLOWING DATA PAIRS (A,B)
A = NUMBER TRADITIONAL NAVAJO HOGANS IN A GIVEN DISTRICT
B = NUMBER MODERN HOUSES IN A GIVEN DISTRICT
THE DATA IS PAIRED BY DISTRICTS ON THE NAVAJO RESERVATION
A RANDOM SAMPLE OF 8 DISTRICTS WERE USED. LET D = B-A
AND LET MU = POPULATION MEAN OF D VALUES.

SOURCE: NAVAJO ARCHITECTURE, FORMS, HISTORY, DISTRIBUTIONS
BY S.C. JETT AND V.E. SPENCER, UNIV. AZ. PRESS

WE WILL USE N = 8 DATA PAIRS AND ALPHA = 0.01
NULL HYPOTHESIS H0: MU = 0
ALTERNATE HYPOTHESIS H1: MU >0

(13.0,18.0) (14.0,16.0) (46.0,68.0) (32.0, 9.0) (15.0,11.0)
(47.0,28.0) (17.0,50.0) (18.0,50.0)

DATA DISK FOR MINITAB WORKSHEET NAME: DWELL.mtp
TI-83 name: DWELL.TXT

Temperature in Miami Versus Honolulu (Paired Difference)

IN THE FOLLOWING DATA PAIRS (A,B)
A = AVERAGE MONTHLY TEMPERATURE IN MIAMI
B = AVERAGE MONTHLY TEMPERATURE IN HONOLULU
THE DATA IS PAIRED BY MONTH STARTING WITH JANUARY
LET D = B-A AND MU = POPULATION MEAN OF ALL D VALUES

SOURCE: STATISTICAL ABSTRACT OF THE UNITED STATES

WE WILL USE N = 12 WITH ALPHA = 0.05
NULL HYPOTHESIS H0: MU = 0
ALTERNATE HYPOTHESIS H1: MU <> 0

(67.5,74.4) (68.0,72.6) (71.3,73.3) (74.9,74.7) (78.0,76.2)
(80.9,78.0) (82.2,79.1) (82.7,79.8) (81.6,79.5) (77.8,78.4)
(72.3,76.1) (68.5,73.7)

DATA DISK FOR MINITAB WORKSHEET NAME: TEMP.mtp
TI-83 name: TEMP.TXT

LIST PRICE VERSUS BEST PRICE FOR A NEW GMC PICKUP TRUCK

IN THE FOLLOWING DATA PAIRS (X,Y)

X = LIST PRICE (IN $1000) FOR A GMC PICKUP TRUCK

Y = BEST PRICE (IN $1000) FOR A GMC PICKUP TRUCK

SOURCE: CONSUMERS DIGEST, FEBRUARY 1994

(12.400, 11.200)	(14.300, 12.500)	(14.500, 12.700)
(14.900, 13.100)	(16.100, 14.100)	(16.900, 14.800)
(16.500, 14.400)	(15.400, 13.400)	(17.000, 14.900)
(17.900, 15.600)	(18.800, 16.400)	(20.300, 17.700)
(22.400, 19.600)	(19.400, 16.900)	(15.500, 14.000)
(16.700, 14.600)	(17.300, 15.100)	(18.400, 16.100)
(19.200, 16.800)	(17.400, 15.200)	(19.500, 17.000)
(19.700, 17.200)	(21.200, 18.600)	

DATA DISK FOR MINITAB WORKSHEET NAME: TRUCK.mtp
TI-83 name:TRUCK.TXT

CRICKET CHIRPS VERSUS TEMPERATURE

IN THE FOLLOWING DATA PAIRS (X,Y)

X = CHIRPS/SEC FOR THE STRIPED GROUND CRICKET

Y = TEMPERATURE IN DEGREES FAHRENHEIT

SOURCE: THE SONG OF INSECTS BY DR. G.W. PIERCE
HARVARD COLLEGE PRESS

(20.000, 88.600)	(16.000, 71.600)	(19.800, 93.300)
(18.400, 84.300)	(17.100, 80.600)	(15.500, 75.200)
(14.700, 69.700)	(17.100, 82.000)	(15.400, 69.400)
(16.200, 83.300)	(15.000, 79.600)	(17.200, 82.600)
(16.000, 80.600)	(17.000, 83.500)	(14.400, 76.300)

DATA DISK FOR MINITAB WORKSHEET NAME: CRICKET.mtp
TI-83 name: CRICKET.TXT

DIAMETER OF SAND GRANULES VERSUS SLOPE ON A NATURAL OCCURRING OCEAN BEACH

IN THE FOLLOWING DATA PAIRS (X,Y)

X = MEDIAN DIAMETER (MM) OF GRANULES OF SAND

Y = GRADIENT OF BEACH SLOPE IN DEGREES

THE DATA IS FOR NATURALLY OCCURRING OCEAN BEACHES.

SOURCE: PHYSICAL GEOGRAPHY BY A.M. KING
OXFORD PRESS , ENGLAND

(0.170, 0.630)	(0.190, 0.700)	(0.220, 0.820)
(0.235, 0.880)	(0.235, 1.150)	(0.300, 1.500)
(0.350, 4.400)	(0.420, 7.300)	(0.850, 11.300)

DATA DISK FOR MINITAB WORKSHEET NAME: SAND.mtp
TI-83 NAME: SAND.TXT

NATIONAL UNEMPLOYMENT RATE MALE VERSUS FEMALE

IN THE FOLLOWING DATA PAIRS (X,Y)

X = NATIONAL UNEMPLOYMENT RATE FOR ADULT MALES

Y = NATIONAL UNEMPLOYMENT RATE FOR ADULT FEMALES

SOURCE: STATISTICAL ABSTRACT OF THE UNITED STATES

(2.900, 4.000)	(6.700, 7.400)	(4.900, 5.000)
(7.900, 7.200)	(9.800, 7.900)	(6.900, 6.100)
(6.100, 6.000)	(6.200, 5.800)	(6.000, 5.200)
(5.100, 4.200)	(4.700, 4.000)	(4.400, 4.400)
(5.800, 5.200)		

DATA DISK FOR MINITAB WORKSHEET NAME: NUNEMPL.mtp
TI-83 NAME: NUNEMPL.TXT

Section 10.5 Antelope Study (Table10-16)

<<<< THUNDER BASIN ANTELOPE STUDY >>>>

COLUMN #1: X1 = SPRING FAWN COUNT/100
COLUMN #2: X2 = SIZE OF ADULT ANTELOPE POPULATION/100
COLUMN #3: X3 = ANNUAL PRECIPITATION (INCHES)
COLUMN #4: X4 = WINTER SEVERITY INDEX (1=MILD , 5=SEVERE)

WE HAVE N = 4 VARIABLES WITH M = 8 DATA POINTS (YEARS)

***** DATA *****

2.90	9.20	13.20	2.00
2.40	8.70	11.50	3.00
2.00	7.20	10.80	4.00
2.30	8.50	12.30	2.00
3.20	9.60	12.60	3.00
1.90	6.80	10.60	5.00
3.40	9.70	14.10	1.00
2.10	7.90	11.20	3.00

DATA DISK FOR MINITAB WORKSHEET NAME: ANTELOPE.mtp
TI-83 NAME: ANTELOPE.TXT

Chapter 10 Using Technology: U.S. Economy Case Study (Table 10-17)

<<<< U.S. ECONOMIC DATA 1976 -TO- 1987 >>>>
COLUMN #1: X1 = DOLLARS/BARREL CRUDE OIL
COLUMN #2: X2 = % INTEREST ON TEN YR. U.S. TRES. NOTES
COLUMN #3: X3 = FOREIGN INVESTMENTS/BILLIONS OF DOLLARS
COLUMN #4: X4 = DOW JONES INDUSTRIAL AVERAGE
COLUMN #5: X5 = GROSS NATIONAL PRODUCT/BILLIONS OF DOLLARS
COLUMN #6: X6 = PURCHASING POWER U.S. DOLLAR (1983 BASE)
COLUMN #7: X7 = CONSUMER DEBT/BILLIONS OF DOLLARS

***** DATA *****

10.90	7.61	31.00	974.90	1718.00	1.76	234.40
12.00	7.42	35.00	894.60	1918.00	1.65	263.80
12.50	8.41	42.00	820.20	2164.00	1.53	308.30
17.70	9.44	54.00	844.40	2418.00	1.38	347.50
28.10	11.46	83.00	891.40	2732.00	1.22	349.40
35.60	13.91	109.00	932.90	3053.00	1.10	366.60
31.80	13.00	125.00	884.40	3166.00	1.03	381.10
29.00	11.11	137.00	1190.30	3406.00	1.00	430.40
28.60	12.44	165.00	1178.50	3772.00	0.96	511.80
26.80	10.62	185.00	1328.20	4015.00	0.93	592.40
14.60	7.68	209.00	1792.80	4240.00	0.91	646.10
17.90	8.38	244.00	2276.00	4527.00	0.88	685.50

DATA DISK FOR MINITAB WORKSHEET NAME: USECON.mtp
TI-83 NAME; USECON.TXT

Section 10.5 problem #3 (Systolic Blood Pressure Data)

<<<< SYSTOLIC BLOOD PRESSURE, SEC. 10.5 PROBLEM #3 >>>>

COLUMN #1: X1 = SYSTOLIC BLOOD PRESSURE
COLUMN #2: X2 = AGE IN YEARS
COLUMN #3: X3 = WEIGHT IN POUNDS

***** DATA *****

132.00	52.00	173.00
143.00	59.00	184.00
153.00	67.00	194.00
162.00	73.00	211.00
154.00	64.00	196.00
168.00	74.00	220.00
137.00	54.00	188.00
149.00	61.00	188.00
159.00	65.00	207.00
128.00	46.00	167.00
166.00	72.00	217.00

DATA DISK FOR MINITAB WORKSHEET NAME: C10S5P3.mtp
TI-83 NAME: C10S5PE.TXT

Section 10.5 problem #4 (Test Scores for General Psychology)

<<<< PSYCHOLOGY TESTS: SEE SECTION 10.5 , PROBLEM #4 >>>>

COLUMN #1: X1 = SCORE ON EXAM #1
COLUMN #2: X2 = SCORE ON EXAM #2
COLUMN #3: X3 = SCORE ON EXAM #3
COLUMN #4: X4 = SCORE ON FINAL EXAM

***** DATA *****

73.00	80.00	75.00	152.00
93.00	88.00	93.00	185.00
89.00	91.00	90.00	180.00
96.00	98.00	100.00	196.00
73.00	66.00	70.00	142.00
53.00	46.00	55.00	101.00
69.00	74.00	77.00	149.00
47.00	56.00	60.00	115.00
87.00	79.00	90.00	175.00
79.00	70.00	88.00	164.00
69.00	70.00	73.00	141.00
70.00	65.00	74.00	141.00
93.00	95.00	91.00	184.00
79.00	80.00	73.00	152.00
70.00	73.00	78.00	148.00
93.00	89.00	96.00	192.00
78.00	75.00	68.00	147.00
81.00	90.00	93.00	183.00
88.00	92.00	86.00	177.00
78.00	83.00	77.00	159.00
82.00	86.00	90.00	177.00
86.00	82.00	89.00	175.00
78.00	83.00	85.00	175.00
76.00	83.00	71.00	149.00
96.00	93.00	95.00	192.00

DATA DISK FOR MINITAB WORKSHEET NAME: C10S5P3.mtp
TI-83 NAME:C10S5P3.TXT

Section 10.5 problem #5 (Hollywood Movies data)

<<<< MOVIE DATA: SEE SECTION 10.5, PROBLEM #5 >>>>

COLUMN #1: X1 = FIRST YEAR BOX OFFICE RECEIPTS/MILLIONS
COLUMN #2: X2 = TOTAL PRODUCTION COSTS/MILLIONS
COLUMN #3: X3 = TOTAL PROMOTIONAL COSTS/MILLIONS
COLUMN #4: X4 = TOTAL BOOK SALES/MILLIONS

***** DATA *****

85.10	8.50	5.10	4.70
106.30	12.90	5.80	8.80
50.20	5.20	2.10	15.10
130.60	10.70	8.40	12.20
54.80	3.10	2.90	10.60
30.30	3.50	1.20	3.50
79.40	9.20	3.70	9.70
91.00	9.00	7.60	5.90
135.40	15.10	7.70	20.80
89.30	10.20	4.50	7.90

DATA DISK FOR MINITAB WORKSHEET NAME: C10S5P5.mtp
TI-83 NAME: C10S5P5.TXT

Section 10.5 problem #6 (All Greens Franchise Data)

<<<< ALL GREENS FRANCHISE DATA: SEE SEC. 10.5 , PROB. #6 >>>>

COLUMN #1: X1 = ANNUAL NET SALES/$1000
COLUMN #2: X2 = NUMBER SQ. FT./1000
COLUMN #3: X3 = INVENTORY/$1000
COLUMN #4: X4 = AMOUNT SPENT ON ADVERTIZING/$1000
COLUMN #5: X5 = SIZE OF SALES DISTRICT/1000 FAMILIES
COLUMN #6: X6 = NUMBER OF COMPETING STORES IN DISTRICT

***** DATA *****

231.00	3.00	294.00	8.20	8.20	11.00
156.00	2.20	232.00	6.90	4.10	12.00
10.00	0.50	149.00	3.00	4.30	15.00
519.00	5.50	600.00	12.00	16.10	1.00
437.00	4.40	567.00	10.60	14.10	5.00
487.00	4.80	571.00	11.80	12.70	4.00
299.00	3.10	512.00	8.10	10.10	10.00
195.00	2.50	347.00	7.70	8.40	12.00
20.00	1.20	212.00	3.30	2.10	15.00
68.00	0.60	102.00	4.90	4.70	8.00
570.00	5.40	788.00	17.40	12.30	1.00
428.00	4.20	577.00	10.50	14.00	7.00
464.00	4.70	535.00	11.30	15.00	3.00
15.00	0.60	163.00	2.50	2.50	14.00
65.00	1.20	168.00	4.70	3.30	11.00
98.00	1.60	151.00	4.60	2.70	10.00
398.00	4.30	342.00	5.50	16.00	4.00
161.00	2.60	196.00	7.20	6.30	13.00
397.00	3.80	453.00	10.40	13.90	7.00
497.00	5.30	518.00	11.50	16.30	1.00
528.00	5.60	615.00	12.30	16.00	0.00
99.00	0.80	278.00	2.80	6.50	14.00
0.50	1.10	142.00	3.10	1.60	12.00
347.00	3.60	461.00	9.60	11.30	6.00
341.00	3.50	382.00	9.80	11.50	5.00
507.00	5.10	590.00	12.00	15.70	0.00
400.00	8.60	517.00	7.00	12.00	8.00

DATA DISK FOR MINITAB WORKSHEET NAME: C10S5P6.mtp
TI-83 NAME: C10S5P6.TXT

MINITAB DATA DISK WORKSHEET CRIME.mtp

THIS IS A CASE STUDY OF EDUCATION, CRIME, AND POLICE FUNDING FOR SMALL CITIES IN TEN EASTERN AND SOUTH EASTERN STATES. THE STATES ARE NEW HAMPSHIRE, CONNECTICUT, RHODE ISLAND, MAINE, NEW YORK, VIRGINIA, NORTH CAROLINA, SOUTH CAROLINA, GEORGIA, AND FLORIDA. THE DATA IS FOR A SAMPLE OF 50 SMALL CITIES IN THESE STATES.

X1 = TOTAL OVERALL REPORTED CRIME RATE PER 1MILLION RESIDENTS
X2 = REPORTED VIOLENT CRIME RATE PER 100,000 RESIDENTS
X3 = ANNUAL POLICE FUNDING IN DOLLARS PER RESIDENT
X4 = PERCENT OF PEOPLE 25 YEARS AND OLDER THAT HAVE HAD 4 YEARS OF HIGH SCHOOL
X5 = PERCENT OF 16 TO 19 YEAR-OLDS NOT IN HIGHSCHOOL AND NOT HIGHSCHOOL GRADUATES.
X6 = PERCENT OF 18 TO 24 YEAR-OLDS ENROLLED IN COLLEGE
X7 = PERCENT OF PEOPLE 25 YEARS AND OLDER WITH AT LEAST 4 YEARS OF COLLEGE

SOURCE: *LIFE IN AMERICA'S SMALL CITIES,* A BOOK BY G.S. THOMAS

X1	X2	X3	X4	X5	X6	X7
478	184	40	74	11	31	20
494	213	32	72	11	43	18
643	347	57	70	18	16	16
341	565	31	71	11	25	19
773	327	67	72	9	29	24
603	260	25	68	8	32	15
484	325	34	68	12	24	14
546	102	33	62	13	28	11
424	38	36	69	7	25	12
548	226	31	66	9	58	15
506	137	35	60	13	21	9
819	369	30	81	4	77	36
541	109	44	66	9	37	12
491	809	32	67	11	37	16
514	29	30	65	12	35	11
371	245	16	64	10	42	14
457	118	29	64	12	21	10
437	148	36	62	7	81	27
570	387	30	59	15	31	16
432	98	23	56	15	50	15
619	608	33	46	22	24	8
357	218	35	54	14	27	13
623	254	38	54	20	22	11

CRIME.mtp continued

X1	X2	X3	X4	X5	X6	X7
547	697	44	45	26	18	8
792	827	28	57	12	23	11
799	693	35	57	9	60	18
439	448	31	61	19	14	12
867	942	39	52	17	31	10
912	1017	27	44	21	24	9
462	216	36	43	18	23	8
859	673	38	48	19	22	10
805	989	46	57	14	25	12
652	630	29	47	19	25	9
776	404	32	50	19	21	9
919	692	39	48	16	32	11
732	1517	44	49	13	31	14
657	879	33	72	13	13	22
1419	631	43	59	14	21	13
989	1375	22	49	9	46	13
821	1139	30	54	13	27	12
1740	3545	86	62	22	18	15
815	706	30	47	17	39	11
760	451	32	45	34	15	10
936	433	43	48	26	23	12
863	601	20	69	23	7	12
783	1024	55	42	23	23	11
715	457	44	49	18	30	12
1504	1441	37	57	15	35	13
1324	1022	82	72	22	15	16
940	1244	66	67	26	18	16

MINITAB DATA DISK WORKSHEET HEALTH.mtp
TI-83 NAME: HEALTH.TXT
THIS IS A CASE STUDY OF PUBLIC HEALTH, INCOME, AND POPULATION DENSITY FOR SMALL CITIES IN EIGHT MIDWESTERN STATES: OHIO, INDIANA, ILLINOIS, IOWA, MISSOURI, NEBRASKA, KANSAS, AND OKLAHOMA. THE DATA IS FOR A SAMPLE OF 53 SMALL CITIES IN THESE STATES.

X1 = DEATH RATE PER 1000 RESIDENTS
X2 = DOCTOR AVAILABILITY PER 100,000 RESIDENTS
X3 = HOSPITAL AVAILABILITY PER 100,000 RESIDENTS
X4 = ANNUAL PER CAPITA INCOME IN THOUSANDS OF DOLLARS
X5 = POPULATION DENSITY PEOPLE PER SQUARE MILE

SOURCE: *LIFE IN AMERICA'S SMALL CITIES*, A BOOK BY G.S. THOMAS

X1	X2	X3	X4	X5
8.0	78	284	9.1	109
9.3	68	433	8.7	144
7.5	70	739	7.2	113
8.9	96	1792	8.9	97
10.2	74	477	8.3	206
8.3	111	362	10.9	124
8.8	77	671	10.0	152
8.8	168	636	9.1	162
10.7	82	329	8.7	150
11.7	89	634	7.6	134
8.5	149	631	10.8	292
8.3	60	257	9.5	108
8.2	96	284	8.8	111
7.9	83	603	9.5	182
10.3	130	686	8.7	129
7.4	145	345	11.2	158
9.6	112	1357	9.7	186
9.3	131	544	9.6	177
10.6	80	205	9.1	127
9.7	130	1264	9.2	179
11.6	140	688	8.3	80
8.1	154	354	8.4	103
9.8	118	1632	9.4	101
7.4	94	348	9.8	117
9.4	119	370	10.4	88
11.2	153	648	9.9	78
9.1	116	366	9.2	102
10.5	97	540	10.3	95

HEALTH.mtp continued

X1	X2	X3	X4	X5
11.9	176	680	8.9	80
8.4	75	345	9.6	92
5.0	134	525	10.3	126
9.8	161	870	10.4	108
9.8	111	669	9.7	77
10.8	114	452	9.6	60
10.1	142	430	10.7	71
10.9	238	822	10.3	86
9.2	78	190	10.7	93
8.3	196	867	9.6	106
7.3	125	969	10.5	162
9.4	82	499	7.7	95
9.4	125	925	10.2	91
9.8	129	353	9.9	52
3.6	84	288	8.4	110
8.4	183	718	10.4	69
10.8	119	540	9.2	57
10.1	180	668	13.0	106
9.0	82	347	8.8	40
10.0	71	345	9.2	50
11.3	118	463	7.8	35
11.3	121	728	8.2	86
12.8	68	383	7.4	57
10.0	112	316	10.4	57
6.7	109	388	8.9	94